Laying the Foundation for Space Solar Power

An Assessment of NASA's Space Solar Power Investment Strategy

Committee for the Assessment of NASA's Space
Solar Power Investment Strategy

Aeronautics and Space Engineering Board

Division on Engineering and Physical Sciences

National Research Council

NATIONAL ACADEMY PRESS
Washington, D.C.

NATIONAL ACADEMY PRESS   2101 Constitution Avenue, N.W.   Washington, DC 20418

NOTICE: The project that is the subject of this report was approved by the Governing Board of the National Research Council, whose members are drawn from the councils of the National Academy of Sciences, the National Academy of Engineering, and the Institute of Medicine. The members of the committee responsible for the report were chosen for their special competences and with regard for appropriate balance.

This study was supported by Contract No. NASW-99037, Task Order 105, between the National Academy of Sciences and the National Aeronautics and Space Administration. Any opinions, findings, conclusions, or recommendations expressed in this publication are those of the author(s) and do not necessarily reflect the view of the organizations or agencies that provided support for the project.

International Standard Book Number: 0-309-07597-1

Available in limited supply from:

Aeronautics and Space Engineering Board
HA 292
2101 Constitution Avenue, N.W.
Washington, DC 20418
(202) 334-2855

Additional copies are available from:

National Academy Press
Box 285
2101 Constitution Ave., N.W.
Washington, DC 20055
(800) 624-6242
(202) 334-3313 (in the Washington metropolitan area)
http://www.nas.edu

Copyright 2001 by the National Academy of Sciences. All rights reserved.

Printed in the United States of America

THE NATIONAL ACADEMIES

National Academy of Sciences
National Academy of Engineering
Institute of Medicine
National Research Council

The **National Academy of Sciences** is a private, nonprofit, self-perpetuating society of distinguished scholars engaged in scientific and engineering research, dedicated to the furtherance of science and technology and to their use for the general welfare. Upon the authority of the charter granted to it by the Congress in 1863, the Academy has a mandate that requires it to advise the federal government on scientific and technical matters. Dr. Bruce M. Alberts is president of the National Academy of Sciences.

The **National Academy of Engineering** was established in 1964, under the charter of the National Academy of Sciences, as a parallel organization of outstanding engineers. It is autonomous in its administration and in the selection of its members, sharing with the National Academy of Sciences the responsibility for advising the federal government. The National Academy of Engineering also sponsors engineering programs aimed at meeting national needs, encourages education and research, and recognizes the superior achievements of engineers. Dr. Wm. A. Wulf is president of the National Academy of Engineering.

The **Institute of Medicine** was established in 1970 by the National Academy of Sciences to secure the services of eminent members of appropriate professions in the examination of policy matters pertaining to the health of the public. The Institute acts under the responsibility given to the National Academy of Sciences by its congressional charter to be an adviser to the federal government and, upon its own initiative, to identify issues of medical care, research, and education. Dr. Kenneth I. Shine is president of the Institute of Medicine.

The **National Research Council** was organized by the National Academy of Sciences in 1916 to associate the broad community of science and technology with the Academy's purposes of furthering knowledge and advising the federal government. Functioning in accordance with general policies determined by the Academy, the Council has become the principal operating agency of both the National Academy of Sciences and the National Academy of Engineering in providing services to the government, the public, and the scientific and engineering communities. The Council is administered jointly by both Academies and the Institute of Medicine. Dr. Bruce M. Alberts and Dr. Wm. A. Wulf are chairman and vice chairman, respectively, of the National Research Council.

COMMITTEE FOR THE ASSESSMENT OF NASA'S SPACE SOLAR POWER INVESTMENT STRATEGY

RICHARD J. SCHWARTZ, *Chair*, Purdue University, West Lafayette, Indiana
MARY L. BOWDEN, University of Maryland, College Park
HUBERT P. DAVIS, Consultant, Canyon Lake, Texas
RICHARD L. KLINE, United Satellite Launch Services, Great Falls, Virginia
MOLLY K. MACAULEY, Resources for the Future, Inc., Washington, D.C.
LEE D. PETERSON, University of Colorado, Boulder
KITT C. REINHARDT, Air Force Research Laboratory, Albuquerque, New Mexico
R. RHOADS STEPHENSON, Jet Propulsion Laboratory (retired), La Canada, California

Liaison from the Aeronautics and Space Engineering Board

DAVA J. NEWMAN, Massachusetts Institute of Technology, Cambridge

Staff

KAREN E. HARWELL, Study Director, Aeronautics and Space Engineering Board
LEE SNAPP, NASA Administrator's Fellowship Program—NRC Visiting Fellow, Aeronautics and Space Engineering Board
GEORGE M. LEVIN, Director, Aeronautics and Space Engineering Board
MARVIN WEEKS, Senior Project Assistant, Aeronautics and Space Engineering Board (May 2000 through March 2001)
MARY LOU AQUILO, Senior Project Assistant, Aeronautics and Space Engineering Board (March through June 2001)
ANNA FARRAR, Administrative Associate, Aeronautics and Space Engineering Board

AERONAUTICS AND SPACE ENGINEERING BOARD

WILLIAM W. HOOVER, *Chair*, U.S. Air Force (retired), Williamsburg, Virginia
A. DWIGHT ABBOTT, Aerospace Corporation (retired), Los Angeles, California
RUZENA K. BAJSCY, NAE, IOM, University of Pennsylvania, Philadelphia
WILLIAM F. BALLHAUS, JR., NAE, Aerospace Corporation, Los Angeles, California
JAMES A. BLACKWELL, Lockheed Martin Corporation (retired), Marietta, Georgia
ANTHONY J. BRODERICK, aviation safety consultant, Catlett, Virginia
DONALD L. CROMER, U.S. Air Force (retired), Lompoc, California
ROBERT A. DAVIS, The Boeing Company (retired), Seattle, Washington
JOSEPH FULLER, JR., Futron Corporation, Bethesda, Maryland
RICHARD GOLASZEWSKI, GRA Inc., Jenkintown, Pennsylvania
JAMES M. GUYETTE, Rolls-Royce North America, Reston, Virginia
FREDERICK H. HAUCK, AXA Space, Bethesda, Maryland
JOHN L. JUNKINS, NAE, Texas A&M University, College Station, Texas
JOHN K. LAUBER, Airbus Industrie of North America, Washington, D.C.
GEORGE K. MUELLNER, The Boeing Company, Seal Beach, California
DAVA J. NEWMAN, Massachusetts Institute of Technology, Cambridge
JAMES G. O'CONNOR, NAE, Pratt & Whitney (retired), Coventry, Connecticut
MALCOLM R. O'NEILL, Lockheed Martin Corporation, Bethesda, Maryland
CYNTHIA SAMUELSON, Logistics Management Institute, McLean, Virginia
WINSTON E. SCOTT, Florida State University, Tallahassee
KATHRYN C. THORNTON, University of Virginia, Charlottesville
ROBERT E. WHITEHEAD, National Aeronautics and Space Administration (retired), Henrico, North Carolina
DIANNE S. WILEY, The Boeing Company, Los Alamitos, California
THOMAS L. WILLIAMS, Northrop Grumman, El Segundo, California

Staff

GEORGE LEVIN, Director

Preface

In 1968, Peter Glaser advanced the proposition that solar energy could be collected by Earth-orbiting satellites and then beamed by means of microwaves to power stations on Earth's surface. The energy collected would be converted to electricity and introduced into commercial power grids for use by customers. Both the Department of Energy and the National Aeronautics and Space Administration (NASA) examined the concept in the late 1970s and early 1980s; however, the program was canceled. In 1995, NASA decided to take a fresh look at the feasibility, technologies, costs, markets, and international public attitudes regarding space solar power (SSP). This *Fresh Look* study[1] found that much had changed. Key technologies needed for the construction, deployment, and maintenance of SSP satellites, such as composite materials, modular fabrication, and robotics for construction and repair, had all shown significant advances. During this period, public concerns about environmental degradation grew. The committee also noted that such environmental concerns are, if anything, even more intense today than in the days of the *Fresh Look* study. As a result of this study, the U.S. Congress became interested in SSP and in FY 1999 appropriated funds for NASA to conduct the SSP Exploratory Research and Technology (SERT) program. The SERT program and its follow-on, the SSP Research and Technology (SSP R&T) program, constitute the effort assessed in this report.[2]

In March 2000, NASA's Office of Space Flight asked the Aeronautics and Space Engineering Board of the National Research Council to perform an independent assessment of the space solar power program's technology investment strategy to determine its technical soundness and its contribution to the roadmap that NASA has developed for this program.[3] The program's investment strategy was to be evaluated in the context

---

[1] Feingold, Harvey, Michael Stancati, Alan Freidlander, Mark Jacobs, Doug Comstock, Carissa Christensen, Gregg Maryniak, Scott Rix, and John Mankins. 1997. *Space Solar Power: A Fresh Look at the Feasibility of Generating Solar Power in Space for Use on Earth.* Report No. SAIC-97/1005. Chicago, Ill.: Science Applications International Corporation (SAIC).

[2] The SERT program was established in FY 1999 and continued through FY 2000 by U.S. congressional appropriation. An additional appropriation was also funded for SSP Research and Technology (SSP R&T) for FY 2001. Decisions on internal NASA budget allocations for FY 2002 were pending during the preparation and review of this report. As a result of recent agencywide realignments, future SSP programs may be included within other NASA initiatives. Throughout this report the term "SERT program" or "SERT effort" refers to both the 2-year Space Solar Power Exploratory Research and Technology (SERT) program during FY 1999 and 2000 and the follow-on effort in FY 2001, referred to as the SSP Research and Technology (SSP R&T) program. The terms "SSP program" and "SSP effort" refer to any planned future program in SSP technology development and are used in recommendations to NASA.

[3] This assessment evaluates the SERT program and the follow-on SSP R&T efforts through December 15, 2000. Program changes after that date are not included.

of its likely effectiveness in meeting the program's technical and economic objectives. *The scope of this study did not include assessments of the desirability of space-generated terrestrial electrical power or assessment of the ability of NASA's space launch development efforts to provide the capability needed to deploy a space solar power system.*

The Committee for the Assessment of NASA's Space Solar Power Investment Strategy of the National Research Council has completed an approximately 12-month study evaluating the technology investment strategy of NASA for SSP. A copy of the statement of task for this study is included in Appendix A. In conducting its review, the committee was not asked to assess, and it did not comment on, the ultimate economic viability of producing terrestrial solar power from space. The committee sees the wisdom of investing some of this nation's resources in a number of potential approaches for dealing with future energy needs. This is particularly true when the committee considers the potential payoffs from this investment to other NASA, government, and commercial programs. This report provides an assessment of NASA's management of its SSP investments and provides recommendations on how its technical investment process can be improved.

The committee recognized that NASA deliberately excluded "lowering the cost of access to space" (i.e., development of new Earth-to-low-Earth-orbit launch vehicles) in its roadmap for SSP technology development. The committee understands and accepts NASA's rationale for this decision. NASA has a major program devoted to lowering the cost of access to space. Given the relatively small amount of funding earmarked by Congress for space solar power technology development, little could be accomplished (and much would be lost) by using these program resources to help lower the cost of access to space.

This study was sponsored by NASA and conducted by a committee appointed by the National Research Council (see Appendix B). The statement of task directed the committee to (1) evaluate NASA's SSP efforts and (2) provide an assessment of its particular investment strategy for a potential program in SSP technology research and development. In order to effectively prioritize and balance investments across several technology areas, rigorous modeling and system analysis studies are usually performed. NASA began this process during the SERT effort. Preliminary technology and programmatic investments were presented to the committee based on this modeling and seem re-

alistic, taking into account the level of funding made available to the program. The committee believes that this approach is one useful technique for assigning technology investment priorities and determining the relative payoff from technology investments. The committee discovered during its meetings, however, that many of the modeling inputs were suspect and that more refinement and better validation were necessary before additional decisions were made regarding technology investment balance. Consequently, the committee agreed that it would be inappropriate to evaluate the actual *magnitude* of funding in each technical area. Comments on the *relative* amounts for various technologies are included.

As a result of low overall program funds during the past 3 years, the program has been forced to make much smaller investments than desired for research in various technical areas. Due to this mismatch between the actual funding and program plan, the committee believed it was critical to evaluate the organizational foundations, modeling methodologies, and program management style on which the future SSP investment strategy will be based (despite levels of funding available to the program). These issues led the committee to perform a two-part assessment of the program, providing (1) an evaluation of the total program investment strategy, management, and organization and (2) an evaluation of each individual SSP-related technology area. The structure of the following report is based on these two factors.

The committee was not asked to evaluate technology development in SSP-related areas in the United States or worldwide or to evaluate any other NASA or non-NASA programs in technology development, whether related to SSP or not. As a result, no other technology program structure was assessed or mentioned in the report. However, knowledge of the state of the art in various technical areas is necessary to effectively evaluate any research and technology effort. Various options for generating power from space have been suggested (and researched) during the past 30 years, including the Lunar Solar Satellite Concept proposed by David Criswell, among others. The committee did not consider such competing concepts for solar power from space but concentrated solely on the NASA SERT program. To this extent, the committee has focused on the program at NASA and its relationship with industry and other efforts in SSP-related technology.

This report has been reviewed in draft form by individuals chosen for their diverse perspectives and technical expertise, in accordance with procedures approved by the National Research Council's (NRC's) Report Review Committee. The purpose of this independent review is to provide candid and critical comments that will assist the institution in making its published report as sound as possible and to ensure that the report meets institutional standards for objectivity, evidence, and responsiveness to the study charge. The review comments and draft manuscript remain confidential to protect the integrity of the deliberative process. We wish to thank the following individuals for their review of this report:

Minoru S. Araki, Lockheed Martin Corporation, retired,
Richard Green, International Power and Environmental Company,
Joel Greenberg, Princeton Synergetics, Inc.,
Nasser Karam, Spectralab, Inc.,
Thomas J. Kelly, Grumman Corporation, retired,
Leeka I. Kheifets, Electric Power Research Institute (EPRI),
Mark S. Lake, Composite Technology Development, Inc.,
F. Robert Naka, CERA, Inc., and
Stephen M. Rock, Stanford University.

Although the reviewers listed above have provided many constructive comments and suggestions, they were not asked to endorse the conclusions or recommendations, nor did they see the final draft of the report before its release. The review of this report was overseen by Gerald L. Kulcinski, University of Wisconsin, appointed by the NRC's Report Review Committee, who was responsible for making certain that an independent examination of this report was carried out in accordance with institutional procedures and that all review comments were carefully considered. Responsibility for the final content of this report rests entirely with the authoring committee and the institution.

The committee also thanks those who took the time to participate in committee meetings and provide background materials (see Appendix E). The committee is especially indebted to Karen Harwell, study director, for her unflagging support of the committee and her help every step of the way. Lee Snapp, a NASA Administrator's Fellowship Program visiting fellow, contributed to the introduction and international sections of the report, and George Levin, director, Aeronautics and Space Engineering Board, was particularly helpful in interpreting and clarifying the committee's charge.

Richard J. Schwartz, *Chair*
Committee for the Assessment of NASA's
Space Solar Power Investment Strategy

Contents

EXECUTIVE SUMMARY 1

1 INTRODUCTION 9
   1-1 Electricity and Solar Power, 9
   1-2 Background, 10
   1-3 Study Approach, 11
   References, 11

2 OVERALL SERT PROGRAM EVALUATION 12
   2-1 Evaluation of Total Program Plan and Investment Strategy, 12
   2-2 Applications, 23
   2-3 International Efforts, 25
   References, 27

3 INDIVIDUAL TECHNOLOGY INVESTMENT EVALUATIONS 29
   3-1 Systems Integration, 29
   3-2 Solar Power Generation, 34
   3-3 Space Power Management and Distribution, 37
   3-4 Wireless Power Transmission, 40
   3-5 Ground Power Management and Distribution, 42
   3-6 Space Assembly, Maintenance, and Servicing, 43
   3-7 Structures, Materials, and Controls, 47
   3-8 Thermal Materials and Management, 51
   3-9 Space Transportation and Infrastructure, 53
   3-10 Environmental, Health, and Safety Factors, 56
   3-11 Platform Systems, 60
   References, 60

APPENDIXES

A	Statement of Task	65
B	Biographical Sketches of Committee Members	66
C	Example of NASA's SERT Program Technology Roadmaps	69
D	Brief Overview of NASA's Space Solar Power Program	73
E	Participants in Committee Meetings	77
F	Acronyms and Abbreviations	79

Tables and Figures

TABLES

ES-1   Proposed Space Solar Power Program Resources Allocation,
       FY 2002 to FY 2006, 2

2-1   NASA's SERT Program—Model System Category Definitions, 13
2-2   Proposed Space Solar Power Program Resources Allocation,
       FY 2002 to FY 2006, 17

D-1   Proposed Space Solar Power Program Resources Allocation,
       FY 2000 to FY 2016, 76

FIGURES

ES-1   Key recommendations to the NASA SSP program, 3

2-1   NASA's SERT program: research and technology schedule of milestones
       roadmap, 14
2-2   NASA's SERT program: strategic research and technology goals, 16

3-1   NASA's SERT program integration process, 30

C-1   Sample SERT progam executive summary chart on solar power generation
       activity, 70
C-2   Sample SERT program goal chart on solar power generation activity, 71
C-3   Sample SERT program milestones chart on solar power generation activity, 72

D-1   Generic space solar power system, 74
D-2   Generic microwave and laser SSP systems, 74

Executive Summary

NASA'S SPACE SOLAR POWER EXPLORATORY RESEARCH AND TECHNOLOGY (SERT) PROGRAM

The National Aeronautics and Space Administration's Space Solar Power (SSP) Exploratory Research and Technology (SERT) program[1] was charged to develop technologies needed to provide cost-competitive ground baseload electrical power[2] from space-based solar energy converters. In addition, during its 2-year tenure, the SERT program was also expected to provide a roadmap of research and technology investment to enhance other space, military, and commercial applications such as satellites operating with improved power supplies, free-flying technology platforms, space propulsion technology, and techniques for planetary surface exploration.

NASA focused the SERT effort[3] by utilizing the definition of a "strawman" or baseline SSP system that would provide 10 to 100 GW to the ground electrical power grid with a series of 1.2-GW satellites in geosynchronous Earth orbit (GEO). For each of the major SSP subsystems, NASA managers developed top-level cost targets in cents per kilowatt-hour (kW-hr) that they felt would have to be met to deliver baseload power at a target of 5 cents/kW-hr. The result of this work was a set of time-phased plans with associated cost estimates that provided the basis for a technology investment strategy. Central to the SERT program was a series of five or six experimental flight demonstrations of progressively larger power-generation capacity, called Model System Categories. These demonstrations will serve as focal points for the advancement of SSP-related technologies and will provide advancements in technologies benefiting other nearer-term military, space, and commercial applications. NASA made extensive use of cost and performance modeling to guide its technology investment strategy.

---

[1]The SERT program was established in FY 1999 and continued through FY 2000 by U.S. congressional appropriation. An additional appropriation was also funded for SSP Research and Technology (SSP R&T) for FY 2001. Decisions on internal NASA budget allocations for FY 2002 were pending during review and publication of this report. During recent agencywide realignments, future SSP programs may be included within other NASA initiatives.

[2]Baseload power is defined as the power available to an area at a constant level during a 24-hour period. For example, most of the power available to residential and business areas is considered baseload power.

[3]Throughout this report the terms "SERT program" and "SERT effort" refer to both the 2-year Space Solar Power Exploratory Research and Technology (SERT) program during FY 1999 and 2000 and the follow-on effort in FY 2001, the SSP Research and Technology (SSP R&T) program. The terms "SSP program" and "SSP effort" refer to any planned future program in SSP technology development.

COMMITTEE ASSESSMENT

The current SSP technology program[4] is directed at technical areas that have important commercial, civil, and military applications for the nation. A dedicated NASA team, operating with a minimal budget, has defined a potentially valuable program—one that will require significantly higher funding levels and programmatic stability to attain the aggressive performance, mass, and cost goals that are required for terrestrial baseload power generation. Nevertheless, significant breakthroughs will be required to achieve the final goal of cost-competitive terrestrial baseload power. The ultimate success of the terrestrial power application depends critically on dramatic reductions in the cost of transportation from Earth to GEO. Funding plans developed during SERT are reasonable, at least during the 5 years prior to the first flight demonstration in 2006 (see Table ES-1). The committee is concerned, however, that the investment strategy may be based on modeling efforts and individual cost, mass, and technology performance goals that may guide management toward poor investment decisions. Modeling efforts should be strengthened and goals subjected to additional peer review before further investment decisions are made. Furthermore, SERT goals could be accomplished sooner and potentially at less cost through an aggressive effort by the SERT program to capitalize on technology advances made by organizations outside NASA.

COMMITTEE RECOMMENDATIONS

Recommendations to the NASA SSP program can be generally categorized by three main imperatives: (1) improving technical management processes, (2) sharpening the technology development focus, and (3) capitalizing on other work. Figure ES-1 provides a snapshot of the committee's key recommendations. Each recommendation is numbered to correspond to the text section in which it is discussed.

Improving Technical Management Processes

NASA's SERT program's technical management processes need to be improved. Currently the program

---

[4]This assessment evaluates the SERT program and the follow-on SSP R&T efforts through December 15, 2000. Program changes after that date are not included.

TABLE ES-1 Proposed Space Solar Power Program Resources Allocation, FY 2002 to FY 2006 (millions of dollars)

Investment Area	FY 2002	FY 2003	FY 2004	FY 2005	FY 2006
Systems integration, analysis, and modeling	5	7	8	8	8
Total technology development	73	92	128	149	154
Technology flight demonstrations	10	25	75	125	150
Total investment	88	124	211	282	312

SOURCE: Adapted in part from "Strategic Research and Technology Road Map." Briefing by John Mankins and Joe Howell, National Aeronautics and Space Administration, to the Committee for the Assessment of NASA's Space Solar Power Investment Strategy, National Research Council, Washington, D.C., December 14, 2000.

has developed a set of integrated roadmaps containing goals, lists of technology challenges and objectives, and a strawman schedule of program milestones that guide technology investment. Appendix C contains a sample set of roadmaps that have been developed for the entire SERT program and each of the program's 12 individual technology areas. The roadmaps' performance, mass, and cost goals are tied to research and technology initiatives in various technical areas necessary for SSP. Unfortunately, the committee did not find adequate traceability between the goals at the system level and those at the subsystem level.

Integral to the milestone schedule are a series of downselect opportunities that precede each flight test demonstration. However, there is no formal mechanism at this point in the program to guide these downselect decisions. The committee has also seen evidence that the current SERT program's roadmaps do not adequately incorporate the planned advances in low-cost space transportation, both Earth-to-orbit and in-space options. Since advancements in space transportation are key to the SSP program's ultimate success, the timing and achievement of technology advances and cost and mass goals by the separate space transportation programs within NASA should be included directly in the SSP roadmaps. A periodic revamping of the roadmaps should be done based on the achievements of NASA in space transportation. SSP

Improve Technical Management Processes

Improve Decision Making
- Written Technology Plan (Rec. 3-1-1)
- Consistent Processes (Rec. 3-1-1)
- Rigorous Systems and Cost Modeling (Rec. 2-1-1)
- Increased Use of Expert Critique/Review (Rec. 3-1-4)

Improve Program Organization
- Continued Use of Flight Test Demonstrations (Rec. 2-1-4)
- Improvement of Advisory Structure (Rec. 3-1-5)

Address Environmental, Health, and Safety Issues Early in Program (Rec. 3-10-1, Rec. 3-10-2)

Sharpen the Technology Development Focus

Invest in Key Enabling Technologies (Rec. 2-1-7)
- Solar Power Generation
- Wireless Power Transmission
- Space Power Management and Distribution
- Assembly, Maintenance, and Servicing
- In-Space Transportation

Under current funding conditions focus on nearer-term applications but maintain the long-term investments in technology development. (Rec. 2-2-1)

Expand Systems Integration and Testing Efforts
- Increased Investment in Modeling Capabilities (Rec. 2-1-1)
- Component-to-System Integration (Rec. 3-1-3)
- Prediction of In-Space Performance (Rec. 3-1-3)

Continue and Expand Technology Demonstrations
- Continued Use of Model System Categories (Rec. 2-1-4)
- Inclusion of Complementary Ground Testing (Rec. 2-1-6)
- Use of International Space Station as Technology Test Bed (Rec. 2-1-5)
- Testing of Robotics and Assembly Techniques on All Flight Test Demonstrations, as Appropriate (Rec. 3-6-2)

Capitalize on Other Work

Earth-to-orbit Transportation (Rec. 3-9-4)

Orbit-to-orbit Transportation (Rec. 3-9-5, Rec. 3-9-6)

Power Generation and Conversion (Rec. 2-1-8, Rec. 3-2-2)

International Efforts (Rec. 2-3-1)

FIGURE ES-1 Key recommendations to the NASA SSP program. Each recommendation is numbered to correspond to the text section in which it is discussed.

program technology investments, flight test demonstrations, and full-scale deployment should be rescheduled accordingly.

Recommendation: NASA's SSP program should improve its organizational and decision-making approach by drawing up a written technology development plan with specific goals, dates, and procedures for carrying out technology advancement, systems integration, and flight demonstration. The SSP program should also establish a consistent process to adjudicate competing objectives within the program and specifically include timing and achievement of technology advances in robotics and space transportation in the roadmaps.

NASA's use of an architecture cost goal estimate based on power costs in the future electricity market is appropriate and commended. As SSP development progresses, however, the architecture cost goal should be adjusted periodically to reflect changes in expectations about future power markets, environmental costs, and other social costs that may arise during development.

The NASA SERT program began development of rigorous modeling and system analysis studies, which were used as a basis for technology and programmatic investments. The approach could be developed, with improvements, into *one* useful technique for determining program priorities. The committee discovered during its meetings that many of the modeling inputs were suspect and that more refinement and better validation were necessary.

Recommendation: The SSP program should review its technology and modeling assumptions, subject them to peer review, and modify where indicated. A single SSP concept should be rigorously modeled, incorporating technology readiness levels and involving industry in conceptual design, as a means to improve the credibility of the model input and output but *not* to prematurely select a single system for ultimate implementation.

The SERT program's oversight advisory structure (called the Senior Management Oversight Committee) includes representatives from various internal NASA organizations, industry, and academia. Further leveraging of technology expertise, management expertise, and funding could be obtained by including representatives from other organizations as well. Additional input would be beneficial from traditional aerospace companies, the Electric Power Research Institute (EPRI), utilities, and other government agencies (particularly the Department of Defense [DOD]/U.S. Air Force, the Department of Energy [DOE], and the National Reconnaissance Office [NRO]). These additions will provide periodic input on the investment strategy and program roadmap and provide further opportunities to validate technological and economic input into the performance and cost models. Individual research and technology working groups have also been established to address planning and technology development in specific technology areas. It would be beneficial to expand these activities.

Recommendation: The current SSP advisory structure should be strengthened with industry (including EPRI and electric utility) representatives plus experts from other government agencies (particularly DOD/Air Force, DOE, and NRO) in order to validate technological and economic inputs into the performance and cost models. Also, due to the wide breadth of technologies related to SSP, the program should establish similar advisory committees for specific technologies in addition to the research and technology working groups currently utilized by the program.

In designing a full-scale SSP system, an environmental impact analysis must be performed that considers human health issues, environmental impact both on Earth and in space, and possible risks to the SSP system itself. Currently the SERT program has placed only a small priority on this area. However, the committee believes that environmental, health, and safety issues should be considered with more emphasis early in the program.

Recommendation: The SSP program should expand its environmental, health, and safety team in order to review SSP design standards (beam intensity, launch guidelines, and end-of-life policies); assess possible environmental, health, and safety hazards of the design; identify research if these hazards are not fully understood; and consider legal and global issues of SSP (spectrum allocation, orbital space, etc.). One approach would be to involve an international organization such as the International Astro-

nautical Federation Space Power Committee in such studies.

Recommendation: Public awareness and education outreach should be initiated during the earliest phases of an SSP program to gain public acceptance and enthusiasm and to ensure ongoing support through program completion.

Sharpening the Technology Development Focus

Key Technologies for SSP

The SERT program must focus its technology development. Currently, the program is funding research in a myriad of technologies that may have potential use in a full-scale SSP system. This research is a valuable endeavor in advancing SSP-related technologies and in determining the extent of development necessary for individual technologies to reach technology readiness levels that can be certified for space flight. *The committee recommends that the current long-term focus of the program remain.* However, due to current funding levels, most investments in individual technologies are much smaller than SERT program managers feel are necessary for adequate research and development of SSP technologies. Many investments are in areas where the utility and power industry should be the lead investor. Under *current* funding constraints, most of the investment should be focused on technologies that have nearer-term applications in space or that may be applied to other Earth applications. Specifically, the committee believes that the greatest benefit would be obtained by investing in several key enabling technologies, which include solar power generation; wireless power transmission; space power management and distribution (SPMAD); space assembly, maintenance, and servicing; and in-space transportation. Without substantial advances in these critical areas, a viable, commercial full-scale SSP system that meets NASA's cost goals may be unattainable in the time frame envisioned by the program.

Solar Power Generation Solar power generation is in the midst of an exciting period of advancement. NASA must collaborate with DOD, DOE, and commercial efforts to avoid undue duplication in research and improve overall effectiveness. Successful attainment of the aggressive cost and mass goals that must be met if SSP is to provide commercially competitive terrestrial power will require that NASA focus on high-reward, high-risk solar array research. Cost-competitive SSP terrestrial electric power will require major technology breakthroughs in solar power generation.

Wireless Power Transmission Investments in wireless power transmission will also need to be focused on more specific areas in the near-term time frame. Currently, the program is funding work on several different options, both microwave and laser. As long as budget levels remain modest, NASA should select one of the three proposed microwave options, along with the laser option, for further funding. Because of the potential benefits to nearer-term space applications, investment in the laser option should be aimed at bringing this technology to the same level of maturity as the microwave option. Ground demonstrations of point-to-point wireless power transmission should be conducted. NASA should also study the desirability of ground-to-space and space-to-space demonstrations.

Space Power Management and Distribution SPMAD is a major contributor to the mass and cost of SSP system designs. Significant investment should be made to reduce the mass and cost of the components to be applied in space while increasing their efficiency and maximum operating temperature. Investments should also be made with companies that are experienced in producing power management and distribution (PMAD) and wireless power transmission components and that will one day have the capability to provide high-volume manufacturing at low cost with high performance and high reliability.

Space Assembly, Maintenance, and Servicing As currently envisioned by the SERT program, autonomous robots will accomplish space assembly, maintenance, and servicing. This will require significant advances in the state of the art of robotics. NASA's SSP program should perform additional systems studies directed at determining the optimal mix of humans and machines and to allow for substantial human involvement on the ground and possibly in space. Focused investments in advancing robotics are expected to have benefits well beyond SSP.

In-Space Transportation Space transportation is key to the deployment of any SSP system. In NASA's initial studies, approximately one-half of the system cost was allocated to ground-to-low-Earth-orbit (LEO) transportation. Earth-to-orbit transportation costs and reliability will be crucial to the deployment of any future commercial SSP system. However, ground-to-LEO transportation is covered by the separate NASA Space Launch Initiative (SLI) program and is outside the scope of this assessment. LEO-to-GEO transportation has little funding in other parts of NASA, so it has been included as part of the SERT program. Chemical, electric, and hybrid propulsion systems are under consideration. In-space transportation is a critical technology that should receive significant investment.

Recommendation: The NASA SSP program should invest most heavily in the following key enabling technologies, mainly through high-payoff, high-risk approaches: (1) solar power generation (in collaboration with DOD/USAF and DOE to avoid duplication); (2) wireless power transmission; (3) space power management and distribution; (4) space assembly, maintenance, and servicing; and (5) in-space transportation. The SSP program should not invest research and development funds in ground PMAD technologies, ground-based energy storage, or platform system technologies. Utilities, industry, and other government programs already have significant investments in those areas.

Recommendation: Under current funding constraints, the SSP program should devote a large portion of its efforts to technologies that have nearer-term applications (e.g., low-mass solar arrays) while continuing to develop technology and concepts for long-term terrestrial baseload power applications.

Any long-term, large program such as SSP must strive to maintain a balance between near- and far-term objectives and goals. Significant differences in technology development would occur if either short- or long-term goals are considered most important. The committee has seen this struggle within the SERT program. Long-term progress must be made in many technology areas before space solar power can become economically viable as a full-scale terrestrial baseload power source. However, due to budget realities and the need to prove near-term success, a program must also make contributions to advancing nearer-term technologies that are applicable to many different programs. In several technology areas, the committee sees merit in suggesting that the SSP program, as *currently funded*, invest in next-generation, revolutionary, high-payoff, high-risk concepts. Each of the 11 individually numbered technical sections in Chapter 3 discusses appropriate long-term recommendations for the program.

Systems Integration

Systems integration is commonly applied during the development phase of a product. However, due to the large number of SSP subsystems and their strong interactions with one another, it should be of vital importance during early SSP technology development. NASA allocated a portion of its SERT funding to developing an overall SSP concept and cost model that includes system cost, mass, and performance targets. Although not yet complete or independently validated, this model has been used as a tool to predict delivered baseload power costs, assuming that various technology goals have been achieved. The committee endorses this methodology as *one* useful technique for assigning technology investment priorities and urges that its development continue as an indicator of the relative payoff from technology investments. The model is still coarse at this time, and the scope and detail should be broadened so that cost and mass targets can be accurately allocated down to the component level. It appears to the committee that many of these goals for launch costs and for system mass and cost must be significantly lower than those currently being used by the NASA team if the system is to produce competitive terrestrial power. Sensitivity studies should be an integral part of any large-scale modeling effort in order to quantify the impacts of departures from the nominal input metrics, many of which are simply assumptions for the SERT program at this time. Nominal input metrics should be developed in consultation with acknowledged experts in SSP-related technology fields to assure quality and accuracy of data.

Recommendation: The SSP team should broaden the scope and detail of the system and subsystem modeling (including cost modeling) to provide a more useful estimate of technology payoff. The models should incorporate detailed concept definitions and include increased input from industry and academia in the specification of model metrics. The costs of transportation, assembly, checkout, and maintenance must also be included in all cost comparisons to properly evaluate alternative technology investment options.

Recommendation: The SSP program should review its technology and modeling assumptions, subject them to peer review, and modify where indicated. A single SSP concept should be rigorously modeled, incorporating technology readiness levels and involving industry in conceptual design, as a means to improve the credibility of the model input and output but *not* to prematurely select a single system for ultimate implementation.

Verification of SSP technology and the integration and testing of hardware and software are necessary before deployment of any SSP system. A combination of new modeling techniques and new design methods, which adaptively accommodate errors in predicted performance and function, may be necessary. The committee saw little evidence of the depth of modeling necessary for such complex space platforms but expects that the effort will increase as candidate designs are chosen.

Recommendation: The SSP program should increase investments in developing spacecraft integration and testing so that the performance of SSP satellites can be verified with a minimum of ground or in-space testing. This may include the development of specialized integration, test, and verification methodologies for SSP spacecraft.

Technology Demonstration

A set of technology flight demonstrations (TFDs) is key to NASA's technology demonstration plan for SSP. Use of these TFDs is commended by the committee as an excellent means of testing available technologies before full-scale integration and deployment. Extensive use of ground demonstration milestones was not observed by the committee in the SERT roadmap. Use of ground demonstrations would provide a lower-cost mechanism to test new technologies before flight. Use of currently available in-space testing mechanisms would also be beneficial to any future SSP program. The current infrastructure on the International Space Station (ISS) could provide an excellent platform for technology demonstration activities. However, because the LEO at which ISS is located may be significantly different from the GEO environment in many ways, demonstration plans should include methodologies that account for the differences between these orbits. Additionally, testing of new robotics and assembly techniques should be incorporated into all flight test demonstrations to further test advanced technologies.

Recommendation: The SSP program should continue the use of technology flight demonstrations to provide a clear mechanism for measuring technology advancement and to provide interim opportunities for focused program and technology goals on the path to a full-scale system.

Recommendation: The SSP program should define additional ground demonstration milestones to be conducted prior to the far more expensive flight tests in order to test advanced technologies and system integration issues before planned downselects of flight-demonstration technologies occur.

Recommendation: NASA should seriously consider utilizing the International Space Station as a technology test bed for SSP during the first set of flight demonstration milestones. Such tests would leverage ISS technology and infrastructure, be independent of new advances in space transportation, and provide an opportunity to test autonomous robotic systems.

Recommendation: The SSP program should perform near-term flight demonstrations of robotic assembly techniques, as well as robotic maintenance and servicing operations. Robotics testing should be incorporated into all SSP flight demonstrations, if possible and as applicable.

Capitalizing on Other Work

NASA's SSP program must capitalize on other work. Even if the SSP funding level increases dramatically, the technical challenges faced by NASA's SSP program will require effective utilization of all resources currently being expended on SSP-related technologies in a variety of government agencies (DOD and DOE), commercial entities, and academia, both in the United States and abroad.

This is especially true in reducing the cost of Earth-to-orbit transportation. NASA's SLI is currently working on cost reduction of transportation to LEO. The SSP program must convey program information to the SLI on its transportation cost goals, optimal payload, mass, packaging, launch rate, and reliability requirements and request that a credible plan be defined by SLI to help achieve these goals. In the case of the proposed use of electric propulsion from LEO to GEO, NASA will need to collaborate with and capitalize on the expertise of commercial firms working on electric propulsion and other in-space transportation options.

Recommendation: The SSP program should begin discussions between its management and that of the NASA Space Launch Initiative, so that future milestones and roadmaps for both programs can reinforce one another effectively. This discussion should include specific information on SSP space transportation needs, including cost goals, timelines for deployment, optimal payload mass, packaging requirements, launch rates, and reliability requirements.

Recommendation: The SSP program should encourage expansion of the current in-space transportation program within NASA and interact with its technical planning to ensure that SSP needs and desired schedules are considered.

Recommendation: The SSP program should increase coordination of industry, academic, and other NASA and non-NASA government investments in advanced in-space transportation concepts, particularly in the areas of electric, solar-electric, magnetohydrodynamic, ion, and solar-thermal propulsion.

The components necessary for the ground PMAD subsystem are similar to those used for terrestrial photovoltaic systems. Substantial research and development work is currently supported by the National Center for Photovoltaics, as well as several commercial entities that provide PMAD components for terrestrial photovoltaic applications. In the case of the solar power generation components (i.e., photovoltaics), programs are currently under way in the Air Force to develop high-efficiency, high-specific-power solar cells. The work of the DOE's National Renewable Energy Laboratory in thin-film solar cells will also be important to the program.

Recommendation: NASA should expand its current cooperation with other solar power generation research and technology efforts by developing closer working relationships with the U.S. Air Force photovoltaics program, the National Center for Photovoltaics, industry, and the U.S. government's Space Technology Alliance.

Although it may be beyond the means of any one country to fund the research, development, and implementation of SSP, these tasks could be more achievable with international cooperation, which would allow NASA to profit from the work of experts worldwide as well as to contribute its own expertise.

Recommendation: NASA should develop and implement appropriate mechanisms for cooperating internationally with the research, development, test, and demonstration of SSP technologies, components, and systems.

Many technologies for SSP (and other space missions) are not currently on the critical path for any near-term NASA mission. Hence, little funding is available that can be leveraged by SSP to develop these technologies. Without this leverage, it is unlikely that the SSP program can be the sole funding source for such technologies. Examples of such technologies are free-flying robotic servicers, specific space structures, reusable in-space transportation, and certain improvements in thermal materials and management and in power management and distribution. While it is beyond the purview of this study to specifically recommend funding increases for programs other than the SSP program assessed in this report, the committee believes that such technologies are important to the ultimate success of SSP.

SUMMARY

The committee has examined the SERT program's technical investment strategy and finds that while the technical and economic challenges of providing space solar power for commercially competitive terrestrial electric power will require breakthrough advances in a number of technologies, the SERT program has provided a credible plan for making progress toward this goal. The committee makes a number of suggestions to improve the plan, which encompass three main themes: (1) improving technical management processes, (2) sharpening the technology development focus, and (3) capitalizing on other work. Even if the ultimate goal—to supply cost-competitive terrestrial electric power—is not attained, the technology investments proposed will have many collateral benefits for nearer-term, less-cost-sensitive space applications and for nonspace use of technology advances.

1

Introduction

1-1 ELECTRICITY AND SOLAR POWER

Throughout history, human progress has been fueled by energy. In early ages, wood to cook food, provide heat, and later to smelt metals provided the primary source of energy. By the seventeenth century, coal to heat homes and factories, make iron and steel, and produce steam for the engines of industrial production was a primary energy source. The twentieth century saw the advent of oil, natural gas, nuclear energy, and various renewable forms of energy, as well as the continuing use of coal to fuel humanity's energy needs. The availability of inexpensive energy that can be converted to usable forms has provided the people of the industrialized nations of the world a standard of living that would have been envied by kings of only a few centuries ago. A nation's economic development and standard of living go hand in hand with readily available, useful forms of energy.

Electricity is one such useful form that can be made from readily available energy sources and is used worldwide. Global demand for electricity has risen tremendously in recent years. In 1990, the world used approximately 11 trillion kW-hr of electricity per year—a figure that is projected to be 22 trillion kW-hr by 2020 (EIA, 2000). However, as this global market grows, other issues have come to the public consciousness. Concerns have arisen about the deterioration of Earth's biosphere and potential long-term changes in climate that may result from pollutants such as carbon dioxide exhausted into the air as a result of fossil fuel combustion.

Using sunlight to generate electricity has been discussed for many years as an alternative source and perhaps a way to relieve some of these concerns. In 1968, in a paper published in *Science*, Peter Glaser proposed that solar energy could be collected by earth-orbiting satellites and then beamed to power stations on Earth's surface (Glaser, 1968). The energy collected would be converted to electricity and introduced into the commercial power grid for use by terrestrial customers. Both the Department of Energy (DOE) and the National Aeronautics and Space Administration (NASA) examined the concept in the late 1970s and early 1980s.

NASA found that generating electric power for terrestrial consumer use was not the only potential application for space solar power. Other uses have been postulated, including power transmission to other space vehicles, power generation for lunar and Martian exploration, power for commercial space development such as communications satellites, and as a source of additional power to enhance the capabilities of such on-orbit facilities as the International Space Station (Grey, 2000). Making some or all of these uses of space solar power a reality requires developing, fielding, and making effective use of a number of complex technologies within a constrained budget. The next section provides a brief history leading up to NASA's current

Space Solar Power (SSP) Exploratory Research and Technology (SERT) program.

1-2 BACKGROUND

From 1979 to 1981, the Committee on Solar Power Systems, Environmental Studies Board, of the National Research Council (NRC) evaluated DOE's and NASA's work on SSP from the 1970s (NRC, 1981). The committee was tasked to perform a critical appraisal of this work, including identifying gaps in the DOE/NASA program and examining the results that the DOE/NASA study obtained. The study's conclusions were not favorable for development of a satellite solar power system. The 1981 NRC report concluded that cost was a major prohibitive factor and the necessary technologies were not of the proper maturity. Estimates of the energy outlook at the time did not indicate that SSP would be a cost-competitive source of electrical energy for the next 20 years. The size and complexity of financing and managing the infrastructure that would be necessary would strain the abilities of the United States. International legal, political, and social acceptability caused by such issues as fear of possible health hazards could make SSP difficult or impossible for the United States to achieve. That earlier report concluded that no funds should be committed specifically to *development* of a satellite solar power system during the next decade. Realizing, however, that circumstances could change that would make more advanced satellite solar power systems an option in the more distant future, the 1981 NRC report also recommended vigorous investigation of technologies relevant to satellite solar power systems that were synergistic with the goals of other programs.

In August 1981, the U.S. Office of Technology Assessment (OTA) also published a report on the DOE/NASA efforts that was unfavorable to continued work in SSP (OTA, 1981). According to the OTA report, too little was known about the technical, environmental, or economic aspects to make a sound decision on whether to continue further development and deployment of SSP. Under the circumstances prevailing at the time, OTA concluded that further research would be necessary before any decisions could be made. When the unfavorable assessments, high initial costs, and need for more research, development, and testing were combined with the drop in oil prices that began in 1984, the urgency that drove development of an SSP system largely evaporated, and work essentially stopped. There was little official interest until the mid-1990s.

In 1995, NASA took a fresh look at the feasibility, technologies, costs, markets, and international public attitudes regarding SSP. The *Fresh Look* study, published in 1997, found that much had changed (Feingold et al., 1997). Several promising concepts were identified as alternatives to the original 1979 reference concept. The study showed that great cost savings, for example, could be achieved over the 1979 reference concept by making use of modular, self-deploying units and on-orbit robotic assembly as opposed to the original concept, involving human-occupied, in-space construction bases. Modularization would also permit the use of smaller launch vehicles in place of a two-stage-to-orbit, reusable, heavy-lift launch vehicle that would require unique ground launch infrastructure. The study noted the critical importance of low-cost transportation to orbit and noted further that, although costs were still too high, the technology to lower launch cost to orbit was separately under development in other NASA programs (although it is uncertain if or when those programs will result in a new generation of launch vehicles or what improvements might be provided in terms of performance or cost). The study asserted that technologies and concepts involved in SSP could become more feasible if both government and commercial non-SSP applications were considered. Finally, the study noted that the market for SSP, though global in nature, might be uncertain for some time to come depending on how various nations' policies treated SSP in comparison with other means of generating electricity (Feingold et al., 1997).

As a result of the *Fresh Look* study, both the U.S. Congress and the Office of Management and Budget became interested in SSP once more. NASA conducted a follow-on concept definition study in 1998. The result was funding of $22 million set aside for NASA to conduct the SERT program. In March 2000, the NASA Office of Space Flight (Code M) approached the NRC with a request to evaluate its technology investment strategy in space solar power with a view to determining whether or not the strategy that the agency had adopted would meet the program's technical and economic objectives.

Although the current NRC committee neither advocates nor discourages SSP, it recognizes that significant changes have occurred since 1979 that might make it worthwhile for the United States to invest in either SSP or its component technologies. Improvements have been seen in the efficiency of crystalline photovoltaic and thin-film solar cells. Lighter-weight substrates and blankets have been developed and flown. A 65-kW solar array has been installed successfully on the International Space Station, and wireless power transmission has been the subject of several terrestrial tests. Japanese and Canadian experiments, some of which are discussed later in this report, have shown that small aircraft can be kept aloft by power transmitted via microwaves. The area of robotics, essential to SSP on-orbit assembly, has shown substantial improvements in manipulators, machine vision systems, hand-eye coordination, task planning, and reasoning. Advanced composites are in wider use, and digital control systems are now state of the art.

Although scientific and engineering advances may help make SSP more feasible, the committee noted that public concerns about environmental degradation are, if anything, even more intense than in the days of the *Fresh Look* study.

1-3 STUDY APPROACH

The NRC formed a committee of eight experts with experience in space systems design, engineering, and launch; solar power generation, management, and distribution; on-orbit assembly; robotics; space structures; and economics to independently assess the technical investment strategy of NASA's space solar power program. The full statement of task, found in Appendix A, asked the committee to address areas of the space solar power investment strategy associated with developmental and operational issues, technical feasibility of various aspects of the program, and opportunities for synergy. The committee restricted its efforts to critiquing NASA's technical investment strategy and neither advocated nor discouraged the concept of space solar power. Assessments of (and comparisons with) other space solar power concepts, such as the Lunar Solar Satellite concept proposed by David Criswell, were not performed by the committee. The committee also did not attempt to predict the role that space solar power might play in the future among the many alternatives for generating electricity. The purpose of this assessment was to evaluate the technology investment strategy of the SERT program and provide guidance as to how the program can be most effective in meeting its long-term goals, not to influence those goals. This assessment evaluates the SERT program and the follow-on SSP R&T efforts through December 15, 2000. Program changes after that date are not included.

The committee approached the study by adopting the SERT program's definition of the term "investment strategy," which includes six areas: (1) program division and organization, (2) use of developmental cycles, (3) opportunities for independent review, (4) balance of internal and external investments, (5) use of systems analysis and modeling to define goals, and (6) periodic review of technology roadmaps. This definition then served as an outline for the approach that the committee used during its assessment.

This report focuses on two levels of assessment: (1) an overall evaluation of the technical investment strategy and program organization and (2) evaluation of individual technology subprograms. Chapter 2 examines the overall investment strategy, the investment strategy methodology, program management issues, and opportunities for synergy with other programs. Recommendations and discussion are categorized in three major areas. Chapter 3 provides individual evaluations of 11 of NASA's 12 technical investment areas (economics is included in the overall assessment in Chapter 2). Recommendations called out in the Executive Summary and listed in Figure ES-1 were considered key by the committee. Other recommendations in Chapter 3 were considered important to managers of individual technology areas.

REFERENCES

EIA (Energy Information Administration). 2000. *International Energy Outlook 2000*. Washington, D.C.: U.S. Department of Energy, p. 114.

Feingold, Harvey, Michael Stancati, Alan Freidlander, Mark Jacobs, Doug Comstock, Carissa Christensen, Gregg Maryniak, Scott Rix, and John Mankins. 1997. *Space Solar Power: A Fresh Look at the Feasibility of Generating Solar Power in Space for Use on Earth*. Report No. SAIC-97/1005. Chicago, Ill.: Science Applications International Corporation (SAIC).

Glaser, Peter. 1968. "Power From the Sun: Its Future." *Science*, Vol. 162, No. 3856, November 22, pp. 857-866.

Grey, Jerry. 2000. "The Technical Feasibility of Space Solar Power." Statement to Subcommittee on Space and Aeronautics, Committee on Science, U.S. House of Representatives. September 7.

National Research Council (NRC), Environmental Resources Board. 1981. *Electric Power from Orbit: A Critique of a Satellite Power System*. Washington, D.C.: National Academy Press.

OTA (Office of Technology Assessment). 1981. *Solar Power Satellites*. NTIS No. PB82-108846. Washington, D.C.: U.S. Government Printing Office, p. 3.

2

Overall SERT Program Evaluation

2-1 EVALUATION OF TOTAL PROGRAM PLAN AND INVESTMENT STRATEGY

The Space Solar Power (SSP) Exploratory Research and Technology (SERT) program was evaluated in the context of the "plan's likely effectiveness to meet the program's technical and economic objectives," as stated in the committee's statement of task (see Appendix A). This top-level assessment leads to identification of the most important technology investment options, opportunities for increased synergy with other efforts, assessment of adequacy of available resources, and possible recommendations for changes in the investment strategy to achieve desired objectives. Discussion and recommendations are grouped into three basic areas: (1) improving technical management processes, (2) sharpening the technology development focus, and (3) capitalizing on other work.

Improving Technical Management Processes

Program Organization

The SERT program was charged to develop technologies needed to provide cost-competitive ground baseload electrical power from space-based solar energy converters. In addition, during its 2-year tenure, the SERT program was also expected to provide a roadmap of research and technology investment to enhance other space, military, and commercial applications such as satellites operating with improved power supplies, free-flying technology platforms, space propulsion technology, and techniques for planetary surface exploration.

With such a broad scope it is not surprising that the National Aeronautics and Space Administration (NASA) centers, the Jet Propulsion Laboratory, and industry participants have defined a myriad of technologies that could be developed for the future applications. It should also not be surprising that if NASA's year-to-year expenditure remains at around $10 million or less, the program will be inadequate to meet the identified needs. Funding has been in yearly incremental add-ons by the U.S. Congress and has not been part of the formal NASA operating plan. It is impossible to make efficient progress in technology development when funding and management support are uncertain. However, the current SERT managers have defined a potentially valuable program despite these obstacles.

Central to the SERT program is a series of experimental demonstrations called model system categories (MSCs) that serve as focal points for the technology definition. Table 2-1 outlines these MSCs (Mankins and Howell, 2000a). Top-level schedule and resources for accomplishing the technology development work as defined by NASA are shown in Figure 2-1 (Mankins and Howell, 2000b). The committee endorses this approach to defining flight test demonstration milestones

TABLE 2-1 NASA's SERT Program—Model System Category Definitions

NASA Model System Category	Power Capability	Flight Test Demonstration Options (to be chosen competitively)	Projected Time Frame
MSC 1	~100 kW	Free flyer LEO-to-Earth power beaming research platform Solar power plug in space Cryogenic propellant depot "Mega-commsat" demonstrator	2006-2007
MSC 1.5	~1 MW	GEO-to-Earth solar power satellite (SPS) demonstrator Lunar exploration SPS platform Earth neighborhood transportation system	2011-2012
MSC 3	~10 MW	Free flyer GEO-based SPS demonstration platforms for wireless power transmission, solar power generation, power management and distribution, and solar electric propulsion Interplanetary transportation system	2016-2017
MSC 4	~1 GW	Commercial space full-scale solar power satellite	2021+

SOURCE: Adapted in part from Mankins and Howell, 2000a.

to validate technology advancement. However, the committee also realizes, as does NASA, that the schedule of milestones and roadmap should be reconfigured as research and development for components of the program are realized (or not realized) and new results are obtained. NASA demonstrated during the course of the study that the roadmaps were revamped several times during the first 2 years of the program in response to both internal agency assessments and external peer review. Continued annual and biannual assessment of the roadmaps, schedules, and goals are an inherent part of the program. The committee recommends that future roadmaps, however, contain more transparent information tying together cost, performance, and schedule. The roadmaps should also more visibly demonstrate their reliance on advances in space transportation and robotics that are entirely or largely funded by other programs.

NASA's SERT program presented a concept for reviewing its time-phased plans, which include the incorporation of NASA's strategic plans and goals, information gleaned from independent program and technology assessments, new innovative technology applications, government and commercial application opportunities, and research efforts in other organizations. This iterative process review would be cycled at least annually because strategic research and technology investments must be selected each fiscal year as part of the NASA budget development process (assuming the work becomes part of the overall NASA program).

The committee has also seen evidence that the current SERT program's roadmaps do not adequately incorporate the planned advances of low-cost space transportation development, both Earth-to-low-Earth-orbit (LEO) and in-space options. Because any advancements in space transportation are key to the SSP program's ultimate success, the timing and achievement of technology advances and cost and mass goals by the separate space transportation programs within NASA should be included directly in the SSP roadmaps. A periodic revamping of the roadmaps should be made based on the achievements of NASA in space transportation. SSP program technology investments, flight test demonstrations, and full-scale deployment should be rescheduled accordingly. Adequate contingency plans also need to be developed to be able to react positively to the failure of any flight or ground test demonstration planned by the program.

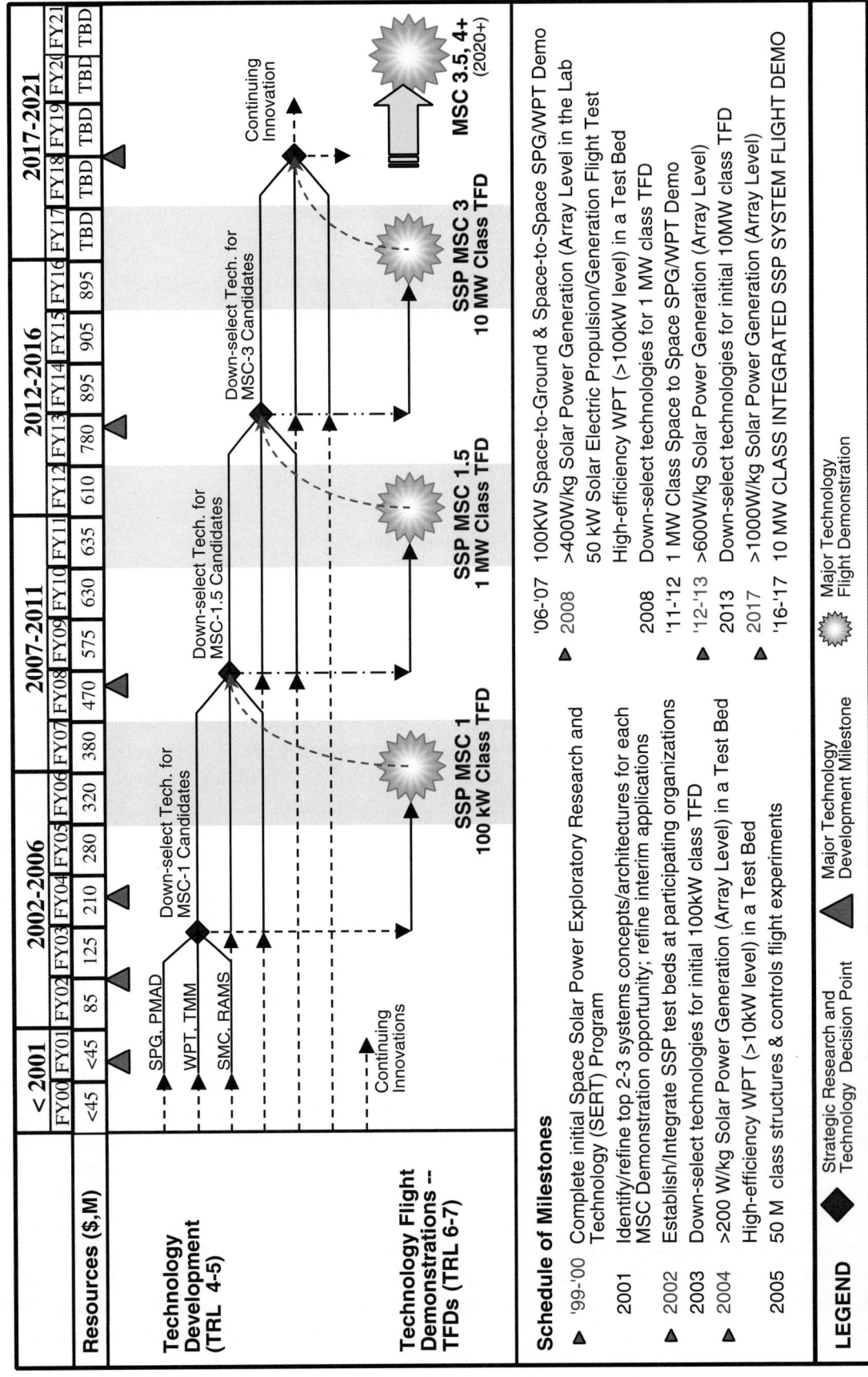

FIGURE 2-1 NASA's SERT program: research and technology schedule of milestones roadmap. NOTE: Figure reprinted in original form. SOURCE: Mankins and Howell, 2000b.

Performance Goals

NASA has made a determined effort in the SERT program to focus the effort by beginning the definition of a "strawman" or baseline SSP system to provide 10-100 GW to the ground electrical power grid with a series of 1.2-GW satellites in geosynchronous Earth orbit (GEO). Since no one knows the time scale to build, launch, and assemble on orbit such a system, the committee did not comment on this particular scenario's potential for commercial appeal (or any of the potential scenarios for MSC 4). Chosen scenarios will be a direct result of the program's investment strategy, the progress of technology development, and a competitive selection process. The committee did not feel it was appropriate to evaluate each individual scenario for a full-scale system. As a result, various scenarios are not presented in the report. Time to market and size of investment necessary for such a system will be issues that need to be addressed as the program progresses; however, the SERT program previously funded an independent economic analysis to evaluate such issues. Assessment of this analysis was outside the scope of this study.

Top-level cost targets in cents per kilowatt-hour were developed for each of the major SSP systems that NASA managers believed were necessary to finally deliver baseload power at less than a selected target of 5 cents/kW-hr.[1] Major system and subsystem functions were each allocated a "contributory" cost goal by program managers. The sum of the contributory goals should, in theory, be equal to the overall cost target of 5 cents/kW-hr. These targets are shown for various design options in Figure 2-2 (Mankins and Howell, 2000b). For brevity, the specific design concepts and options (Mankins and Howell, 2000a; Carrington and Feingold, 2000) listed in Figure 2-2 are not presented in the report. The NASA program plans to continue monitoring this target as markets for electricity change and to adjust this target and its distribution among technologies accordingly. As such, the manner in which *current* cost goals are set is justified.

A corresponding set of mass, cost, and performance targets was then used to help define where technology funds should be applied, and detailed roadmaps have been developed to accomplish these technology goals. The result of this work is a set of time-phased plans with associated cost estimates, which provide the basis for an investment strategy. The committee notes that there is a lack of traceability (of cost and mass goals) to the next lower level. The committee expects that in future program documents there will be traceability of cost and mass targets down to the subsystem level and to the component level. Without consistent cost and mass goals with clear traceability from the top level to the component technology level, individual technology teams may not make the most appropriate technology investments.

The major SERT system cost and performance targets, as shown in Figure 2-2, are extremely aggressive.[2] Additionally, they include reliance that NASA's separate Space Launch Initiative (SLI) program will be successful in reducing Earth-to-LEO transportation costs to $400/kg. NASA's second-generation SLI goal is $2,200/kg, and the third-generation goal is approximately $220/kg (NASA, 1999; Davis, 2000). In a SERT Program Status report (Mankins, 2000b), NASA reported that *current* SERT concepts (December 13, 2000) result in predicted costs for power in the range of 10-20 cents/kW-hr, versus NASA's full-scale system goal of 5 cents/kW-hr.

NASA has adopted an allowable cost of 5 cents/kW-hr as its target goal for competitive terrestrial power production. The committee suggests that this value be revisited as the program proceeds; however, it is viewed by the committee as a reasonable starting point for the investment strategy. This choice sets the revenue stream level for a 1.2-GW facility. Once the revenue stream is known, the net present value of this revenue stream can be computed. A simplified calculation was made by the committee for the required return on investment, assuming zero operating costs and a 40-year operating period. The calculation demonstrates the importance of strengthening the cost analysis for the operational system. For instance, using a 10 percent rate of return, $5 billion is available for the entire system.

---

[1] This 5 cents/kW-hr goal was based on cost estimates gleaned by NASA from an independent economic analysis (Macauley et al., 2000).

[2] Subsystem cost, mass, and performance targets were also supplied to the committee for each technical area in the program's work breakdown structure. For brevity, only the top-level program goals are presented in this publication. More specific information can be obtained from the listed reference.

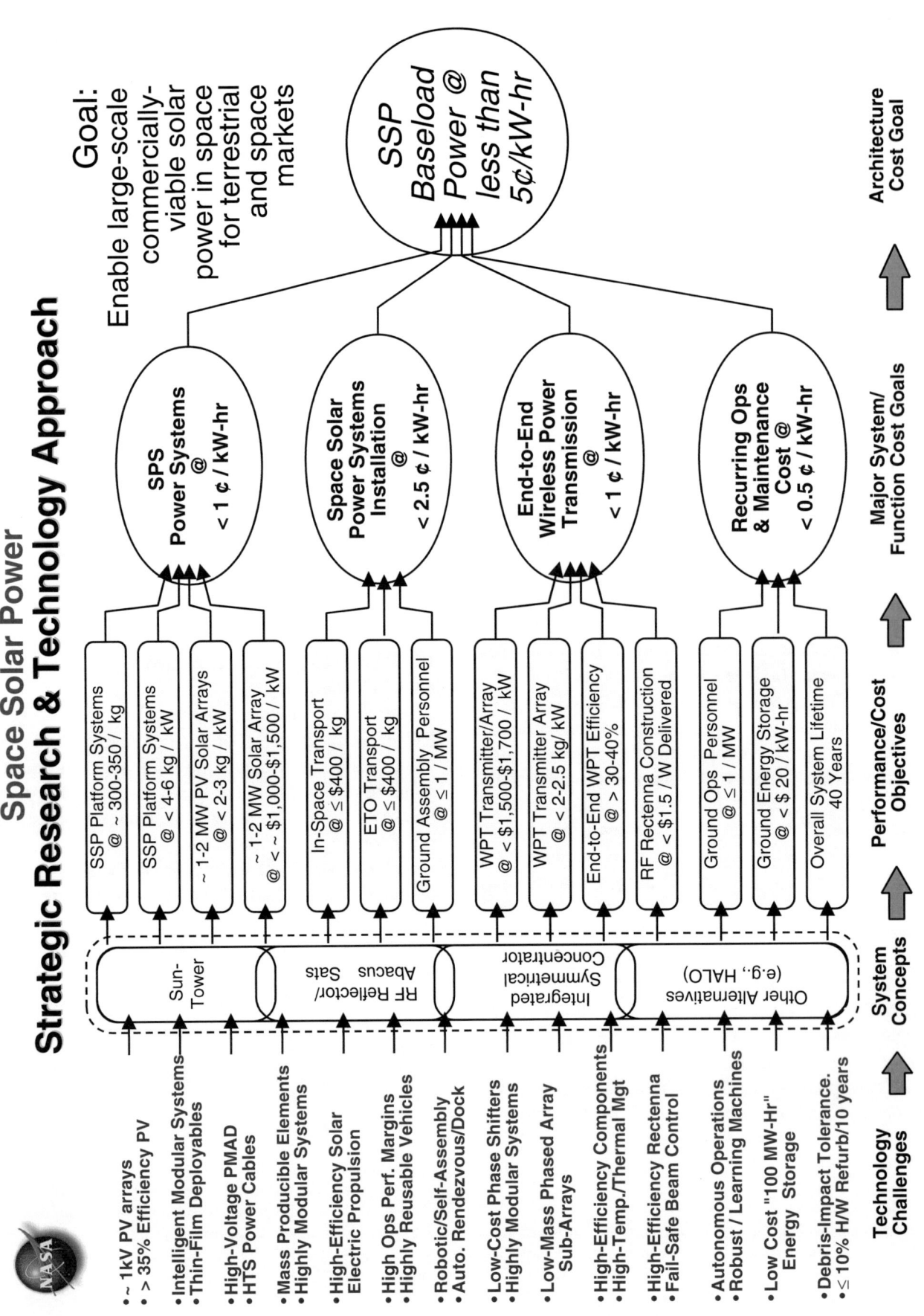

FIGURE 2-2 NASA's SERT program: strategic research and technology goals. NOTE: Figure reprinted in original form. SOURCE: Mankins and Howell, 2000b.

Using the NASA target goal of allocating 2.5 cents of the 5 cent/kW-hr revenue stream to launch costs, in-space transportation, and ground assembly personnel allows $2.5 billion to be spent on construction and installation of the system. At an assumed launch cost of $800/kg to GEO, it would cost $14 billion to launch a 1.2-GW system unless the currently assumed weight of the 1.2-GW facility is substantially reduced. The calculations should be developed further consistent with stationary power plant funding procedures and perhaps using the models for fission nuclear power introduction. In any event, more stringent cost and mass goals must be achieved in order to meet *current* NASA cost goals for competitive terrestrial electric power. The NASA team must be more rigorous in their cost and mass allocations to the subsystems and components, as well as launch costs. Current space transportation and technology subsystem goals, for example, are already driving near-term choices within the SERT program, creating the possibility that technology investment choices may be based on goals that are not stringent enough in this area.

Consequently, the committee believes that NASA not only should reevaluate its cost goals in various technology areas but also should complete a rigorous analysis of its cost goals in the space transportation area. Many of the goals for launch costs and system mass and cost must be significantly lower than currently being used by the NASA team if the system is to produce competitive terrestrial power. Allocations should also be made in absolute terms, dollars and kilograms, as well as familiar ratios such as dollars per watt and watts per kilogram. These absolute terms can be used directly by SSP technical staff in answering the question, How much can this specific technology subsystem cost to attain an overall 5 cents/kW-hr cost? The committee notes that the 5 cents/kW-hr target may be unnecessarily low for nonterrestrial applications of space solar power, such as space-to-space power beaming for interplanetary spacecraft or space-to-planetary surface beaming for rovers, for example.

Resource Allocation

NASA's Space Solar Power Strategic Research and Technology Roadmap proposes resource allocations as given in Table 2-2. The figures are broken down into systems integration, analysis, and modeling, total technology development, and flight test demonstrations (referred to by NASA as technology flight demonstra-

TABLE 2-2 Proposed Space Solar Power Program Resources Allocation, FY 2002 to FY 2006 (millions of dollars)

Investment Area	FY 2002	FY 2003	FY 2004	FY 2005	FY 2006
Systems integration, analysis, and modeling	5	7	8	8	8
Total technology development	73	92	128	149	154
Technology flight demonstrations	10	25	75	125	150
Total investment	88	124	211	282	312

SOURCE: Adapted in part from "Strategic Research and Technology Road Map." Briefing by John Mankins and Joe Howell, National Aeronautics and Space Administration, to the Committee for the Assessment of NASA's Space Solar Power Investment Strategy, National Research Council, Washington, D.C., December 14, 2000.

tions). Breakouts were also provided for each of the main technology development categories providing a description of the proposed work, schedule, and cost goals (see Appendixes C and D for more detailed information). The committee restricted its attention to the first 5 years of the program (FY 2002 to FY 2006) due to the large uncertainty in the out years.

The committee's reactions to NASA's proposed resource allocations are as follows:

- This is a reasonable first projection of resource requirements through FY 2006.
- The committee supports NASA's approach of dividing the program into three major elements: (1) systems integration, analysis, and modeling, (2) technology development, and (3) technology flight demonstration. In the future, if some ground demonstrations are added as key milestones, it would be appropriate to group them with the technology flight demonstrations and rename the category "technology demonstration."
- Although it was not given a cost breakdown for the eight subelements within systems integration, analysis and modeling, the committee would expect the systems and infrastructure modeling to decrease markedly after the first few years and be replaced by technology validation and studies of mission architecture.
- Technology development is broken down by main resource applications. NASA's chosen distribution shows emphasis in the areas of greatest impact: solar power generation (SPG); wireless power trans-

mission (WPT); space power management and distribution (SPMAD); and space assembly, maintenance, and servicing. However, the relative size of the resources should vary much more between the vital technologies and those that serve to make advances on all fronts. Additionally, it appears that this first estimate does not leverage the work of other agencies in such critical areas like SPG and SPMAD or, if it does, that leverage is very small.

- Technology flight demonstration resources may be at an appropriate level for the first flight test demonstration (MSC 1). However, a requirements-driven conceptual study, including cost development, is needed before strong endorsement can be offered. Specifically, the requirements to be validated must be clearly defined, and alternative concepts for meeting them should be developed, including associated costs and schedules. Potential resource support from other agencies and international sources should be defined, resulting in a full conceptual plan that can be critically evaluated.

- The time schedule for each technology flight demonstration should be tied to demonstrating desired technology levels, rather than an arbitrary date.

Sharpening the Technology Development Focus

System and Cost Modeling

A commendable start has been made on systems and cost modeling for various SSP concepts and technology choices (Carrington and Feingold, 2000; Feingold, 2000; Mullins, 2000). However, the committee believes there may still be a great disparity between the resources that can reasonably be expected and the desired rate of technology development progress. This will require careful evaluation of technology payoffs and tough program management decisions.

The SERT program's general approach of defining a baseline concept and coupling it with detailed system and subsystem performance plus cost modeling to guide technology investment will be an excellent tool for making some of these crucial decisions provided the modeling is strengthened, as described in Section 3-1 of this report. Additionally, the technology goals and roadmaps concept should provide a credible goal for individual technology areas. However, for this methodology to be realistic, goals must flow from the technology and cost requirements in both a top-down and bottom-up manner, involving input and decision making from both technology management and technologists.

Conceptually, the use of an architecture cost goal estimate based on power costs in the future electricity market is appropriate and commended. Specification of this goal as a probability distribution representing a range of uncertainty would better represent the uncertainty inherent in any projections of future market potential. Over time, as SSP development progresses, the architecture cost goal should be adjusted to reflect changes in expectations about future power markets, environmental costs, and other social costs that may arise.

The SSP cost and system analysis models should incorporate one detailed concept definition, making it possible to evaluate the payoff of specific technology efforts within the broad functional systems areas of solar power generation and wireless power transmission, among others. This concept definition should also include assembly, checkout, and maintenance techniques. The model needs to clearly show where SSP technology investments branch off to the benefit of other missions, including both in-space and terrestrial applications.

Input should be gathered from hardware builders, industry, academia, and other government agencies to improve the modeling effort. The input should include information on technical performance, current state-of-the-art technology, forecasts for technology advancement, and system cost modeling. The expansion of the modeling should also include other applications (in addition to terrestrial power supply) that will benefit from the SSP technology investments.

Future comparisons in the models should include transportation costs and delivered performance on orbit, which will enhance decision making between alternative technologies. This seems to have been intended in some of the past SERT program trade studies; however, the committee believes that the principle is not sufficiently ingrained in the modeling effort.

Recommendation 2-1-1: The SSP team should broaden the scope and detail of the system and subsystem modeling (including cost modeling) to provide a more useful estimate of technology payoff. The models should incorporate detailed concept definitions and include increased input from indus-

try and academia in the specification of model metrics. The costs of transportation, assembly, checkout, and maintenance must also be included in all cost comparisons to properly evaluate alternative technology investment options.

In the economics modeling of new technologies, taking explicit account of risk and uncertainty is key to an effective understanding of the sensitivity of models to assumptions and the quality of the data used to inform the models. The SERT program is currently using conventional measures of technology readiness (the Technology Readiness Level index [Mankins, 1995]) and a rough measure of technological uncertainty (the Research and Development Degree of Difficulty [Mankins, 1998]) in its modeling effort. These measures address dimensions of the engineering stages of development of new technologies but are silent as to these technologies' ultimate usefulness and cost-effectiveness, which are important factors in setting priorities for roadmaps. The program may be better served by integrating measures in the cost and system modeling that consider the cost-effectiveness of new technologies, the relationships among different technologies, and their effect on overall program structure.

The SERT team began to take this step in results presented to this committee in December 2000 (Mankins, 2000b). Further development would provide a more effective, resilient, and defensible roadmap for SSP, that is, not only the integrated technology analysis methodology but also explicit incorporation of risk, uncertainty, cost, and benefit data. However, subsequent decision making based solely on these models is *not* encouraged by the committee until the models have been adequately validated and tested against baseline SSP concepts (see Section 3-1). In addition to this validation, model assumptions and their possible impacts on cost, use of certain technologies, and mass should also be evaluated before further decisions are made.

It is also useful that the roadmaps clearly distinguish "risk" from "uncertainty"—words often used ambiguously. Risk is generally understood as describing a known probability of an undesirable outcome—failure—while uncertainty refers to lack of knowledge about potential outcomes (Knight, 1921). The assessment of risk thus depends critically on the definition of failure, which in turn may depend on public and institutional (e.g., government agencies, Congress) expectations.

Recommendation 2-1-2: The SSP team should continue integration of technology readiness measurement and cost uncertainty modeling, both in developing a consistent framework for the approach and in parameterizing the framework with the best available information. Over time, the empirical evaluation of these dimensions should become better understood as technology demonstrations begin.

Overall Program Focus in a Cost-Constrained Environment

Under the current NASA funding environment (i.e., yearly congressional earmarks), the program has been provided with funds that are adequate only for technology roadmap development and preliminary planning. With these constraints, it is difficult to fund every technological area with promise for SSP. The committee suggests that if full program funding is not made available, additional focusing of the SSP effort be made. With the projected level of funding, managers should select a single principal baseline concept and one technology per subsystem. For example, a magnetron might be selected as part of the WPT baseline concept. The baseline should be altered and more advanced technologies substituted when justified by technical progress and funding.

Recommendation 2-1-3: The committee recommends additional focusing of the SSP program. For example, if continued underfunding of the effort continues, it would be in the program's best interest to choose a single baseline concept and one technology per subsystem for technology advancement. There should be periodic reviews as technologies are advanced, and alternatives may be reintroduced into the design when justified and affordable.

Technology Demonstration

The committee endorses the approach of defining demonstration milestones of achievement (NASA's MSC categorization) to provide program focus, as well as a clear mechanism for measuring technology advancement progress. Progression from the 100-kW MSC 1 demonstration to more technically challenging and larger demonstration missions (MSC 1.5, 2, 3, etc.) is reasonable. The committee appreciates the fact that the MSC 1 demonstration is not dependent on simultaneous invention of a new low-cost transportation system. The committee also encourages the continued balance of technology flight demonstrations that test near-term and far-term technologies. Advancements in SSP-related technologies will be beneficial to the entire space program. Flight demonstrations currently planned in the program are designed to test the interactions and feasibility of advanced technologies for possible use with SSP systems. These demonstrations will test technologies at technology readiness levels that are below the level industry will accept. Without flight test experiments, industry will be hesitant to accept many of the new technologies into their programs.

Recommendation 2-1-4: The SSP program should continue the use of technology flight demonstrations to provide a clear mechanism for measuring technology advancement and to provide interim opportunities for focused program and technology goals on the path to a full-scale system.

The committee also sees value in testing technologies for SSP on already available space platforms. In particular, it believes that NASA should consider ways in which the International Space Station (ISS) can help advance the technology development effort in SSP. Furthermore, industry and academia should openly compete on proposals for technology demonstration so that their expertise is brought to bear on the technical issues. In addition, international cooperation will most likely be necessary for any large-scale terrestrial SSP application, and international markets for SSP should be considered, for economic reasons.

Flight demonstration of technologies and testing of systems should be completed early in the SSP program. The program's use of flight demonstration milestones is an excellent way of achieving this. Although the committee recognizes the vast difference between conditions and operational procedures in LEO and GEO, early milestone tests may have to be performed at LEO. To save cost and to maximize the engineering data returned, these experiments would have a smaller scope than the MSC system-level demonstrations. They might, for example, involve measurement of a particular mechanical effect of zero gravity on a subscale structural component rather than on a complete, functioning solar array.

Serious consideration should be given to developing the first flight demonstration (MSC 1) as an ISS technology research mission and developing other experimental programs to be validated on the ISS.[3] The NASA SERT program's 100-kW MSC 1 free flyer could be assembled from the ISS as a technology demonstration test bed. For instance, various solar array concepts could be used on MSC 1 and then subjected to test after release from the ISS. Space-to-space transmission could also be demonstrated using a co-orbiting target module, which also could serve as a platform for performing experiments at lower microgravity levels than will be achievable on the ISS. After test completion, MSC 1 could be returned to the ISS for inspection and subsequent reoutfitting with other subsystems for a second round of free-flyer tests. Experiments testing low-power-level transmission to Earth should also be included for microwave and laser wireless power transmission systems—both of which are being considered in NASA's SSP program. Transmission efficiency would be low for these initial tests, especially for microwave systems, but the experimental performance could be corrected analytically to correspond to full-scale systems. An added advantage to ISS-based assembly of the MSC 1 is that various assembly, maintenance, and service techniques could be tested and developed for future application in GEO.

Upon its completion, the MSC 1 demonstration could provide an enhancement for the ISS technology demonstration program. This enhancement could be either use of a free flyer in co-orbit with ISS or an upgrade to the ISS solar array with new technology. When the ISS program is ready to procure a replacement or expansion of the ISS solar arrays, the program should consider the SSP array technology and, if possible, procure an array that is on the roadmap for an SSP system. Additionally, the ISS might be used as an orbiting platform to validate techniques for structural assembly, subsystem life, repair, and other experiments unique to an SSP program.

Opportunities for collaboration with the Department of Defense (DOD) and international technologists should be considered in the MSC 1 flight demonstration program for technology leveraging and shared funding. The Air Force Research Laboratory's

---

[3]One previous NRC study recommended that NASA use the International Space Station as a test bed for engineering research and technology development. Areas suggested that overlap with SSP-related technologies included electric power, robotics, structures, and thermal control (NRC, 1996).

PowerSail program is a 30- to 50-kW thin-film photovoltaic free-flyer demonstration intended to fly in the 2004 or 2005 time frame. This may be an opportunity for collaboration with NASA's MSC 1 demonstration.

Recommendation 2-1-5: NASA should seriously consider utilizing the International Space Station as a technology test bed for SSP during the first set of flight demonstration milestones. Such tests would leverage ISS technology and infrastructure, be independent of new advances in space transportation, and provide an opportunity to test autonomous robotic systems.

The starting point for considering all of these collaborative options is the development of a detailed set of requirements to be met by MSC 1 in support of SSP and other applications. Development of the technology demonstration concept and evaluation of the advantages and disadvantages presented by using the ISS should then follow.

There is a need for a comprehensive ongoing program to advance critical technologies from the laboratory to operational readiness. This program must include focused flight demonstrations for many of the individual technologies, in addition to the system-level MSC flight demonstrations. The committee recommends definition of additional ground demonstration milestones to be conducted prior to the far more expensive flight tests. In addition, the committee feels that each of the demonstration projects should be evaluated against the goals for the project and the timing, based on the technology available at the time.

Recommendation 2-1-6: The SSP program should define additional ground demonstration milestones to be conducted prior to the far more expensive flight tests in order to test advanced technologies and system integration issues before planned downselects of flight-demonstration technologies occur.

Technology Building Blocks

Specific treatment of the SERT technology building blocks can be found in Chapter 3; however, several general observations can be made from NASA's modeling data that influence the investment strategy:

- By any yardstick, current expenditures of $10 million to $40 million per year cannot come close to providing the technology development progress that is necessary for application of terrestrial SSP in the next 20 years.
- Due to uncertainties in future funding for SSP, various near-term choices must be made by SSP program managers. Even with a large increase in funding, NASA's SSP program would be best served by focusing its efforts more narrowly than at present. Most of the technology investments should be devoted to technologies that have multiple applications (in addition to terrestrial power generation). In many of the key enabling SSP technologies, significant advances must be made in technology performance, mass, and cost before a commercial SSP system is viable. Most far-term investments should be in research areas that are high risk but could provide high payoff to the SSP program.
- The SSP program should give considerable weight to nearer-term space, military, and commercial applications of this technology, or portions of it (e.g., low-mass solar arrays or WPT). Only a few technologies unique to terrestrial power generation should be funded by NASA. Specifically, the system studies indicate that greatest benefit is obtained by investing most heavily in several key technologies, described below (Carrington and Feingold, 2000; Feingold, 2000; Mullins, 2000):

  —Solar power generation technology is currently in the midst of an exciting period of advancement with solar array improvements of benefit to all solar-powered applications, including terrestrial power and space vehicles. NASA should collaborate with DOD, DOE, and commercial efforts to avoid duplication and improve the overall effectiveness of investments in SPG technology.

  —Wireless power transmission has possible dual application potential for free-flying platforms in space or airborne, which could attract military or commercial participants. However, investments in this area need to be focused. Currently, the SERT program is funding efforts in several major WPT technologies.

  —Space power management and distribution is a major contributor to SSP system mass and cost. Investments should be made to reduce the mass and cost of the components while increasing efficiency and improving operation conditions. New SPMAD techniques developed under the SSP program will have application to the ISS and many other NASA, DOD, and commercial systems.

  —Space assembly, maintenance, and servicing

are challenges common to all large space systems and planetary exploration, particularly when launching a complete unit is beyond the capacity of the space transportation vehicle. SSP systems should be designed from the outset to accommodate on-orbit robotic assembly and maintenance. The degree of structural deployment versus assembly should be rigorously studied, and deployment concepts should be developed that are compatible with planned robotic capabilities. Systems studies are necessary to determine what level of robotic capability is optimal (from tele-operated to fully autonomous) and how humans are best used in the assembly, maintenance, and servicing operations (on the ground or in orbit).

—In-space transportation is an important driver in establishing on-orbit SSP costs. Trade studies of various concepts should be made along with satellite design and assembly concepts to establish the lowest-cost methods for placing the completed SSP design in GEO.

- Utilities, industry, and other government programs should make the most investment in ground power management and distribution (PMAD) technologies, ground-based energy storage, and platform system technologies. These areas are either utility specific or are funded adequately through other efforts. Furthermore, key research and development should place heavy emphasis on reduction of mass and cost and improvements in efficiency, the ultimate drivers for commercial application of any SSP system.

Recommendation 2-1-7: The NASA SSP program should invest most heavily in the following key enabling technologies, mainly through high-payoff, high-risk approaches: (1) solar power generation (in collaboration with DOD/USAF and DOE to avoid duplication); (2) wireless power transmission; (3) space power management and distribution; (4) space assembly, maintenance, and servicing; and (5) in-space transportation. The SSP program should not invest research and development funds in ground PMAD technologies, ground-based energy storage, or platform system technologies. Utilities, industry, and other government programs already have significant investments in those areas.

Capitalizing On Other Work

It was clear from material presented to the committee that the SSP program, including SSP for terrestrial use and other technology applications, is highly synergistic with the related work of other U.S. agencies, and commercial and international interests. NASA must develop strong interfaces with other organizations to ensure that funds are spent most effectively. Specifically, the U.S. Air Force has a vigorous space photovoltaics technology program that could support NASA and potentially benefit from many of the SSP demonstration programs. There may also be near-term commercial and military applications for WPT to power long-duration airships and aircraft.

Similarly, DOE, with its Office of Energy Research and the National Renewable Energy Laboratory, should be involved, for both near-term benefits and long-term program planning. DOE currently spends approximately $75 million/year on solar power technologies and also supports research in related environmental health and safety areas. Internationally, Europe and Japan are rapidly increasing funding for terrestrial photovoltaics research.

The U.S. government sponsors a Space Technology Alliance to increase coordination and collaboration on space-related technology development. Member agencies include the Air Force Research Laboratory, the Ballistic Missile Defense Organization, the Defense Advanced Research Projects Agency, the National Reconnaissance Office, the Naval Research Laboratory, DOE, and NASA, among others. Current initiatives of its Space Power Subcommittee include infusion of technologies such as photovoltaics, modular low-cost PMAD, and efficient compact thermal management into all government space programs. NASA is currently involved in the alliance; however, continued and increased involvement is suggested by the committee to promote increased technology leveraging.

International coordination should be fostered on environmental, safety, and spectrum allocation issues, as well as on other standard space technology development topics. The committee encourages mutually beneficial cooperation consistent with the International Traffic in Arms Regulations, perhaps with a research cooperative agreement, to maximize the effectiveness of the total investment.

Recommendation 2-1-8: NASA should expand its current cooperation with other solar power generation research and technology efforts by developing closer working relationships with the U.S. Air Force photovoltaics program, the National Center for

Photovoltaics, industry, and the U.S. government's Space Technology Alliance.

2-2 APPLICATIONS

NASA's SSP technologies have extensive applications both in space and terrestrially. The program's technology development can also be leveraged by other internal NASA programs and industry. Technology development crucial for an SSP system includes solar array advancements; robotic maintenance and servicing development; power management and distribution; and enhanced systems integration activity for large technical programs. These technologies have applications to many other engineering and science efforts in both government and industry. The following sections outline NASA's current efforts in these areas and provide recommendations for future activities in relation to potential applications of SSP and areas for potential technology transfer.

Applications to Enable Space Science

The objective of the NASA SERT science effort is to "enable science efforts to identify, focus, and quantify the scientific benefit of new concepts created by providing a source of beamed power in space and the science missions made possible by the SSP developed technologies" (Marzwell, 2000). The basic premise of the effort is that high power and large, lightweight structures can enable in-space science. The higher power available from an SSP system can be used for increased penetrating power for imaging and sounding instruments; increased power for drilling and the volatilization of subsurface materials with beamed energy; and increased mobility and power for remote drillers, rovers, and moles in areas where current power systems cannot provide adequate resources.

Many SSP technologies have already been identified by NASA to enable science (Marzwell, 2000). Advanced space-based structures, including large apertures, large photovoltaic arrays, and space-rigidized aerobrake structures, are expected to be a secondary product of SSP research and development. Laser-electric and solar-thermal propulsion and beamed energy power could be directly applicable to Earth-orbit and Mars-orbit missions as well as to future lunar activities. SSP research and development can also improve active sensing technology utilized to map hidden surfaces of planets and asteroids, discover new planetary bodies, analyze atmospheric properties, perform surface imaging, and track resources such as ice and water. Power beaming has application in space (Earth orbit, Mars orbit), as power from an orbit to a planetary surface, or in transportation (including laser sails and laser-thermal and laser-electric propulsion). SSP can also be utilized as an inexpensive, abundant power source for conventional orbital science. Several specific applications are being pursued by NASA in these areas.

HEDS Applications

The strategic plan of NASA's Human Exploration and Development of Space (HEDS) effort establishes a range of visionary goals and objectives, including multiple targets for human exploration outside LEO, goals for scientific discovery through research in space, and research to enable humans to live and work permanently in space. The SERT program has identified potential space applications of SSP technologies and concepts in relation to the HEDS effort (Mankins, 2000a).

Preliminary assessments of many applications have been performed by NASA. These preliminary assessments are an excellent initiation in providing motivation for the development of a much smaller SSP system in addition to the development of a future terrestrial baseload power system. Assessments were divided into categories based on the requirements of SSP technology. Nearer-term applications such as the use of evolutionary power systems for the ISS and solar power systems for GEO communications satellites are considered Generation I applications. Generation II applications include robotic planetary outpost power systems, wireless planetary power grids, and public space travel and tourism. Industrial space stations in LEO, power plugs in space, in-space propellant depots, and integrated human and robotic exploration applications are considered Generation III and IV applications. These applications would utilize SSP technologies in the mid- to far-term time frame (2005-2025). Many other potential applications such as space business parks, space utilities, interplanetary electromagnetic propulsion, and solar system resource development are very far-term applications. The committee believes that, while these far-term applications are important, the SSP program may be better served by focusing on nearer-term applications and technology under current funding conditions.

Recommendation 2-2-1: Under current funding constraints, the SSP program should devote a large portion of its efforts to technologies that have nearer-term applications (e.g., low-mass solar arrays) while continuing to develop technology and concepts for long-term terrestrial baseload power applications.

These applications for power may provide a better business case for industry development during the commercialization phase because such applications will be able to absorb power costs that are higher than future terrestrial power markets will accept. The committee believes, however, that pursuit of such large-scale applications may be possible only with increased cooperation between NASA and foreign governments and industry. The committee also recognizes that this proposed SSP program will require the development of closer working relationships with industry. NASA should lay the groundwork for commercialization.

Recommendation 2-2-2: The SSP program, as well as any future effort in space solar power on the part of NASA, should involve a concerted effort to develop closer ties to industry, the U.S. government, international groups, and other internal NASA efforts for purposes of technology development, peer review, and possible shared resources.

There is a need to involve more outside people in the Senior Management Oversight Committee (SMOC) or the appropriate technical interchange meetings and system working groups meetings (not just organizations external to NASA that are funded by the program, but others as well). Continued and expanded peer review should be an element of any future SSP effort. This may simply be accomplished by expanding the role of SMOC, including more industry and academic researchers in various areas germane to SSP technology development. These actions on the part of NASA may alleviate the issues raised in other sections of this report on the validity and reality of technical assumptions and forecasts (see Sections 2-1 and 3-1).

Technology Transfer and Cross-Cutting Applications

Individual technologies labeled as SSP-enabling technologies also have potential for terrestrial use and use in other space or aviation activities. Basic research in support of SSP is being performed in the areas of photon interaction and ablative physics, wireless power transmission, photovoltaics, robotics, and advanced materials development. Such technology advancement activities are important endeavors that should be continued. One previous NRC study recommended that NASA use the ISS as a test bed to develop new space technologies (NRC, 1996). Many SSP-enabling technologies, such as robotics, solar arrays, structures, WPT, and assembly techniques, could be tested on the ISS.

Unfortunately, little support has been seen within NASA for cross-enterprise technology programs except from the personnel directly involved in such efforts (NRC, 1998). Increased leverage of current NASA programs directly related to SSP-enabling technologies should be pursued. The NASA SERT program has attempted, on a preliminary basis, to coordinate its research and technology roadmaps with other NASA programs. Specific examples follow:

- HEDS Technology-Commercialization Initiative;
- Gossamer Spacecraft program;
- Small Business and Innovative Research program;
- Spacecraft power and propulsion "core technology competency" funding, i.e., the Cross-Enterprise Technology program;
- Advanced Space Transportation and Space Launch Initiative programs;
- Intelligent Systems (IS) program; and
- Space science spacecraft technology demonstration programs (e.g., New Millennium).

The SERT program has also attempted coordination with various efforts external to NASA, including DOE's National Renewable Energy Laboratory photovoltaics research, the National Science Foundation's efforts in innovative manufacturing and robotics, and DOD and the Naval Research Laboratory's work in the area of intelligent systems. Program managers agree that the SSP effort within NASA needs to improve its track record of coordination with other research and technology programs across NASA, the U.S. government, and non-U.S.-government organizations. Furthermore, there is a need to incorporate flight test demonstrations of key technologies on space platforms such as the International Space Station or geosynchronous communications satellites. Also, as reflected in Section 2-3, the SERT program has had little detailed discussion with international efforts except in the area of frequency allocation. One improvement in international relations seen in the last 2 months of the study was the organization of an Open Forum on Space Solar Power, with participation by several countries. More activities such as this one are encouraged by the committee.

Recommendation 2-2-3: The SSP program should assess in detail how to use research by other organizations to expedite the development of SSP-enabling technologies.

During the tenure of this study, the SERT program did indeed begin what was termed a "gap analysis" in order to uncover areas within NASA in which increased technology leveraging may occur. The assessment of investments external to NASA is scheduled to be completed by the time this report is published, but as of December 2000 efforts had not yet been initiated. As the NASA SERT program management agrees, it will be imperative for the SSP program to carry this analysis a step further by actually using the knowledge from these yet-to-be-identified programs and leading efforts to adjust the focus of other related NASA technology development programs to help achieve SSP research and technology goals. Factors to be considered in such an evaluation should include information solicited from the outside research community and program balance between various issues such as technology push versus program pull, near-term versus far-term applications, and competitive technology development versus the need for system design choices. An example of such integrated technology planning and cross-enterprise technology can be found in NASA's Office of Space Science; it is described in a National Research Council study (NRC, 1998).

As part of the 2-year SERT effort, NASA contracted with the American Institute of Aeronautics and Astronautics (AIAA) to perform an assessment of certain aspects of NASA's SSP research and technology effort (AIAA, 2000). Part II of the report discussed multiple-use technologies and applications. It is expected that these findings will be incorporated into the current SSP effort at NASA. The report identified multiple use of SSP technologies as a major area for in-depth consideration. The prospects for these technologies were evaluated by AIAA in order to answer the following questions:

- How real are these technologies and applications?
- Are there other as-yet-unearthed opportunities for dual or alternative applications of SSP technology?
- How and to what extent should SSP technology studies be integrated with these non-SSP programs?
- What might be the payoffs of such an integration in achieving the various programs goals?

Initial areas for consideration included geocentric space applications, lunar and planetary exploration, space science projects, national security applications in space, terrestrial applications, and a miscellaneous category (AIAA, 2000). Two areas mentioned by the AIAA report but not covered in the NASA material presented to this committee were national security missions and terrestrial applications such as airborne vehicles and offshore oil platforms. These areas, as well as the other applications, should be given greater consideration as potential applications for an SSP system. A need still exists to explore further potential applications and technology transfer opportunities, particularly in relation to the importance of the SSP effort having industrial support throughout the program's lifetime. As previously witnessed during the Apollo and shuttle programs, technology development in computers, materials, and robotics can be transferred directly to many everyday industrial and personal applications.

2-3 INTERNATIONAL EFFORTS

The committee examined several activities in progress outside the United States and noted a growing worldwide interest and involvement in space solar power. With a new global energy market emerging, led by electricity as the fastest-growing form of energy for users worldwide, NASA has excellent opportunities to contribute to and profit from international collaboration. Advantages to worldwide cooperation stem not only from the synergy that is possible from cooperation with other experts but also from the fact that SSP has space-based components and thus no technically imposed geographic limits on the countries that could participate in SSP's benefits (AIAA, 2000). The committee was briefed on current international involvement in SSP and found an optimistic global picture. Japan, France, Canada, Russia, Ukraine, Georgia, Italy, Belgium, Germany, India, Netherlands, China, and

Singapore are among the countries engaged in at least some facets of SSP studies, research, development, and technology demonstration. Several multinational organizations such as the United Nations and the European Space Agency (ESA) are also sponsoring work in SSP (Erb, 2000).

The following sections show the diversity of international efforts in which NASA's participation might be mutually beneficial.

SPS 2000

SPS 2000 is a planned operational test bed for several important space solar power technologies. The project was originally proposed by Makoto Nagatomo of the Institute of Space and Astronautical Science, Sagamihara, Japan (Pignolet, 1999), working with a team of government and academic research scientists. The SPS 2000 proposal features a 10-MW solar power satellite in an 1,100-km equatorial orbit. The 1,100-km orbit corresponds to about a 100-min orbital period, permitting the satellite to furnish a power burst of about 200 s to a rectenna[4] on the ground (Moore, 2000). As of November 30, 2000, 19 sites in 11 equatorial countries had been selected as candidate rectenna sites. Field visits by the SPS 2000 task team have permitted the team to establish local contacts, engage the population in the project, and examine local impacts (Mori, 2000).

The Grand-Bassin Project

The French government, with the Centre National d'Études Spatiales (CNES) as its primary agent, also favors international cooperation, believing that success will come when a critical mass of research that makes use of technical synergy is applied to SSP problems (Vassaux, 1999). To that end, CNES, the University of La Réunion, and researchers from Japan are sponsoring a large-scale wireless power demonstration project on the island of La Réunion. La Réunion is a volcanic island in the Indian Ocean southeast of Madagascar. Because the island has few indigenous conventional energy resources, it must import them at great expense. Local decision makers actively support new ways to provide energy to their population (Pignolet, 1999).

The village of Grand-Bassin on La Réunion Island is located in a deep canyon 700 m distant from an entry point to the island's power grid. The Grand-Bassin project involves transmitting about 10 kW of electricity via microwave energy at 2.45 GHz to provide power for the village. This project is notable for going beyond current experimental feasibility of WPT to an operational system that would be subject to real-world demands: exposure to tropical weather conditions, varying power demand level, continuous power production, and reasonable cost. CNES estimates that the system could be in place as early as 2003 (Pignolet, 1999).

Demonstrations of Wireless Power Transmission

Both Japan and Canada have been involved in demonstrations of WPT. A survey taken in Canada identified some 50 organizations in that country that have interests that relate to SSP. Canadian interests include space-based collection systems, space-to-ground demonstrations, ground point-to-point WPT demonstrations, and pilot plants. Canadian activity includes planning for a ground demonstration site, probably in Newfoundland, that will examine many of the same technical issues that are planned for Grand-Bassin, although in completely different climate and terrain (Erb, 2000). Canada also sponsored development of a prototype microwave-powered aircraft for use as a long-duration, high-altitude communications platform. First flights of a scale model began in 1986 (Erb, 2000).

At Kobe University in Japan in 1995, researchers successfully flew a small airship using power transmitted from the ground by microwave energy. Researchers estimated that they were able to generate 10 kW of radiated power from the parabolic antenna on the ground and obtained about 5 kW from the rectenna onboard the airship. Researchers were able to use the power to make the airship climb as high as 45 m above the ground and to hover there for 4 min 15 s (Kaya, 1999).

Other International Efforts

The committee noted numerous other projects that dealt directly with SSP or with the development of technologies that will facilitate SSP. The Technical Uni-

---

[4]A rectenna (the term was coined by the late William Brown, formerly of Raytheon), sometimes referred to as a rectifying antenna, is composed of a mesh of dipoles and diodes for absorbing microwave energy from a transmitter and converting it into electric power (DC current).

versity of Berlin has been involved in SSP studies since 1985. German work has included computer simulations of GEO solar power satellites that include operational characteristics and life-cycle costs; modeling the acquisition and operation of SSP lunar installations; and modeling the development of SSP in conjunction with human exploration of the Moon and Mars (AIAA, 2000).

In Russia, the Central Science Research Institute of Machine Building has designed an SSP satellite that would have a maximum output of 3.14 MW. This design would transmit solar power by microwave to orbiting or remote spacecraft and to lunar and Martian bases. In Ukraine, researchers are proposing experiments for the International Space Station dealing with semiconductor structures for solar cells, new forms of solar concentrators, and enhanced forms of power transmission (AIAA, 2000).

The nations of the European Space Agency are also currently active in SSP studies and research. Early in 2000, ESA published a study conducted jointly by German, Italian, and French participants that sought to identify opportunities, capabilities, and technologies for exploration and utilization of space in the next 30 years. Among the SSP scenarios examined in this study were an experiment for ISS that involved microwave power transmission to a free-flying satellite, wireless power transmission from one terrestrial location to another via a power relay satellite in GEO, and solar-power-generating satellites for terrestrial power supply (Nordlund, 2001).

NASA's Role in International SSP Activities

It was not the committee's intent to produce an exhaustive list of efforts worldwide but to show that there are opportunities for international collaboration. The committee noted that NASA hosted an International Space Power Forum at NASA Headquarters in Washington, D.C., in January 2001. At that forum, representatives from Canada, the ESA, and Japan all indicated that increased interest and participation by the United States would be beneficial to work they were doing in their own countries. The committee understood that the International Traffic in Arms Regulations may impede transfer of certain technologies but also noted that a great deal can be accomplished under current law and that avenues exist for approving technologies for export.

It may be beyond the means of any one country to fund the research, development, and implementation of SSP, but these tasks should be more achievable by international cooperation. International cooperation would allow NASA to profit from the work of experts worldwide, as well as to contribute its own expertise. The comments at the 2001 International SSP Forum from Canadian, European, and Japanese representatives reinforce the committee's observation that NASA's collaboration internationally would be welcome and appropriate.

Recommendation 2-3-1: NASA should develop and implement appropriate mechanisms for cooperating internationally with the research, development, test, and demonstration of SSP technologies, components, and systems.

REFERENCES

AIAA (American Institute of Aeronautics and Astronautics). 2000. *AIAA Assessment of NASA Studies of Space Solar Power Concepts*. Prepared for National Aeronautics and Space Administration, Office of Space Flight, under NASA Grant NAG8-1619. Reston, Va.: American Institute of Aeronautics and Astronautics.

Carrington, Connie, and Harvey Feingold. 2000. "SERT Systems Integration, Analysis, and Modeling." Briefing by Connie Carrington, National Aeronautics and Space Administration, and Harvey Feingold, SAIC, to the Committee for the Assessment of NASA's Space Solar Power Investment Strategy, National Academy of Sciences, Washington, D.C., September 13.

Davis, Danny. 2000. "2nd Generation RLV Summary." Presentation prepared by Danny Davis, Marshall Space Flight Center, for the Committee for the Assessment of NASA's Space Solar Power Investment Strategy, National Academy of Sciences, Washington, D.C., October 24.

Erb, R.B. 2000. "Interest and Activities in Space Solar Power Outside the USA." Briefing by Bryan Erb, Canadian Space Agency, to the Committee for the Assessment of NASA's Space Solar Power Investment Strategy, National Academy of Sciences, Washington, D.C., December 14.

Feingold, Harvey. 2000. "SERT Systems Integration, Analysis, and Modeling." Briefing by Harvey Feingold, SAIC, to the Committee for the Assessment of NASA's Space Solar Power Investment Strategy, National Academy of Sciences, Washington, D.C., October 23.

Kaya, Nobuyuki. 1999. "Ground Demonstrations and Space Experiments for Microwave Power Transmission." *Space Energy and Transportation* 4(3, 4):117-123.

Knight, Frank. 1921. *Risk and Uncertainty*. New York: Houghton Mifflin.

Macauley, Molly, Joel Darmstadter, John Fini, Joel Greenberg, John Maulbetsch, Michael Schaal, Geoffrey Styles, and James Vedda. 2000. *Can Power from Space Compete? The Future of Electricity Markets and the Competitive Challenge to Satellite Solar Power*. Discussion Paper 00-16. Washington, D.C.: Resources for the Future.

Mankins, John. 1995. *Technology Readiness Levels*. NASA White Paper (April 6). Washington, D.C.: National Aeronautics and Space Administration.

Mankins, John. 1998. *Research and Development Degree of Difficulty*. NASA White Paper (March 10). Washington, D.C.: National Aeronautics and Space Administration.

Mankins, John. 2000a. "Human Exploration and Development of Space—Space Solar Power Applications." Briefing by John C. Mankins to the Committee for the Assessment of NASA's Space Solar Power Investment Strategy, National Academy of Sciences, Washington, D.C., September 14.

Mankins, John. 2000b. "Space Solar Power Exploratory Research and Technology (SERT) Program Status." Briefing by John Mankins, National Aeronautics and Space Administration, to the Committee for the Assessment of NASA's Space Solar Power Investment Strategy, National Academy of Sciences, Irvine, Calif., December 14.

Mankins, John, and Joe Howell. 2000a. "Space Solar Power (SSP) Exploratory Research and Technology (SERT) Program Overview." Briefing by John Mankins and Joe Howell, National Aeronautics and Space Administration, to the Committee for the Assessment of NASA's Space Solar Power Investment Strategy, National Academy of Sciences, Washington, D.C., September 13.

Mankins, John and Joe Howell. 2000b. "Strategic Research and Technology Roadmap." Briefing by John Mankins, National Aeronautics and Space Administration, to the Committee for the Assessment of NASA's Space Solar Power Investment Strategy, National Academy of Sciences, Irvine, Calif., December 14.

Marzwell, Neville. 2000. "Space Science Enabled Applications." Briefing by Neville Marzwell, Jet Propulsion Laboratory, to the Committee for the Assessment of NASA's Space Solar Power Investment Strategy, National Academy of Sciences, Washington, D.C., September 14.

Moore, Taylor. 2000. "Renewed Interest in Space Solar Power." *EPRI Journal* 25(1):6-17.

Mori, Masahiro. 2000. *Summary of Studies on Space Solar Power Systems of the National Space Development Agency of Japan*. White Paper (November 30). Tokyo, Japan: Advanced Mission Research Center, Office of Technology Research, pp. 31, 33.

Mullins, Carie. 2000. "Integrated Architecture Assessment Model and Risk Assessment Methodology." Briefing by Carie Mullins, Futron Corporation, to the Committee for the Assessment of NASA's Space Solar Power Investment Strategy, National Academy of Sciences, Washington, D.C., October 23.

NASA (National Aeronautics and Space Administration). 1999. *Space Launch Initiative Program Description*. NASA White Paper. Washington, D.C.: National Aeronautics and Space Administration. Available online at <http://std.msfc.nasa.gov/sli/aboutsli.html>. Accessed August 15, 2001.

Nordlund, Frederic. 2001. "ESA and Space Solar Power (SSP)." Presentation by Frederic Nordlund, European Space Agency, to the International Space Power Forum, Washington, D.C., January 12.

NRC (National Research Council), Aeronautics and Space Engineering Board. 1996. *Engineering Research and Technology Development on the Space Station*. Washington, D.C.: National Academy Press. Available online at <http://books.nap.edu/catalog/9026.html>. Accessed August 16, 2001.

NRC, Space Studies Board. 1998. *Assessment of Technology Development in NASA's Office of Space Science*. Washington, D.C.: National Academy Press. Available online at <http://www.nationalacademies.org/ssb/tossmenu.htm>. Accessed August 16, 2001.

Pignolet, Guy. 1999. "The SPS-2000 'Attaché Case' Demonstrator." *Space Energy and Transportation* 4(3, 4): 125-126.

Vassaux, Didier. 1999. "The French Policy on Space Solar Power and the Opportunities Offered by the CNES Related Activities." *Space Energy and Transportation* 4(3, 4):133-134.

3

Individual Technology Investment Evaluations

3-1 SYSTEMS INTEGRATION

Systems integration is commonly utilized during the development phase of a product but is a vital consideration in the early space solar power (SSP) technology development work due to the large number of subsystems and their strong interaction with one another. Committee comments and recommendations are grouped into three areas: (1) improving technical management processes, (2) sharpening the technology development focus, and (3) capitalizing on other work.

Improving Technical Management Processes

As discussed in Section 2-1, NASA has allocated some of its available SERT funding to the development of an overall SSP concept and cost model that includes system mass and performance targets (Carrington and Feingold, 2000; Feingold, 2000; Mullins, 2000). This model, while not yet complete or independently validated, has been used as a tool to predict delivered baseload terrestrial power costs assuming various technology goals have been achieved. The committee endorses this methodology as one useful technique for assigning technology investment priorities and urges that its development continue as an indicator of the relative payoff from technology investments. The model is coarse at this time and must be expanded so that cost and mass targets can be allocated down to the lowest (component) level. Sensitivity studies, which the SERT program has begun, should be an integral part of such modeling, to quantify the impacts of departures from the nominal input values, many of which are, at this time, simply assumptions. Nominal values should be developed in consultation with acknowledged experts in the field. The modeling activity should also be extended to other near-term applications of SSP technologies. Integration, however, requires much more than the exercise of a modeling tool. An effective team process is necessary to coordinate actions and to ensure that the best possible decisions are made for SSP.

NASA has established a working process for integrating the SERT effort. This program decision-making and organization process to integrate SERT technology working groups and task forces has been well coordinated and has succeeded in advancing the SERT program definition during the past 2 years. Figure 3-1 provides a schematic of the relationships and interactions among the various groups. The committee notes the exemplary degree to which NASA's SERT project organization has been open and responsive to outside suggestions. That openness indicates a high degree of objectivity and confidence and should be encouraged in all continued work. However, three factors will emerge to force the SERT-type integration process to change in the future: (1) funding levels must increase

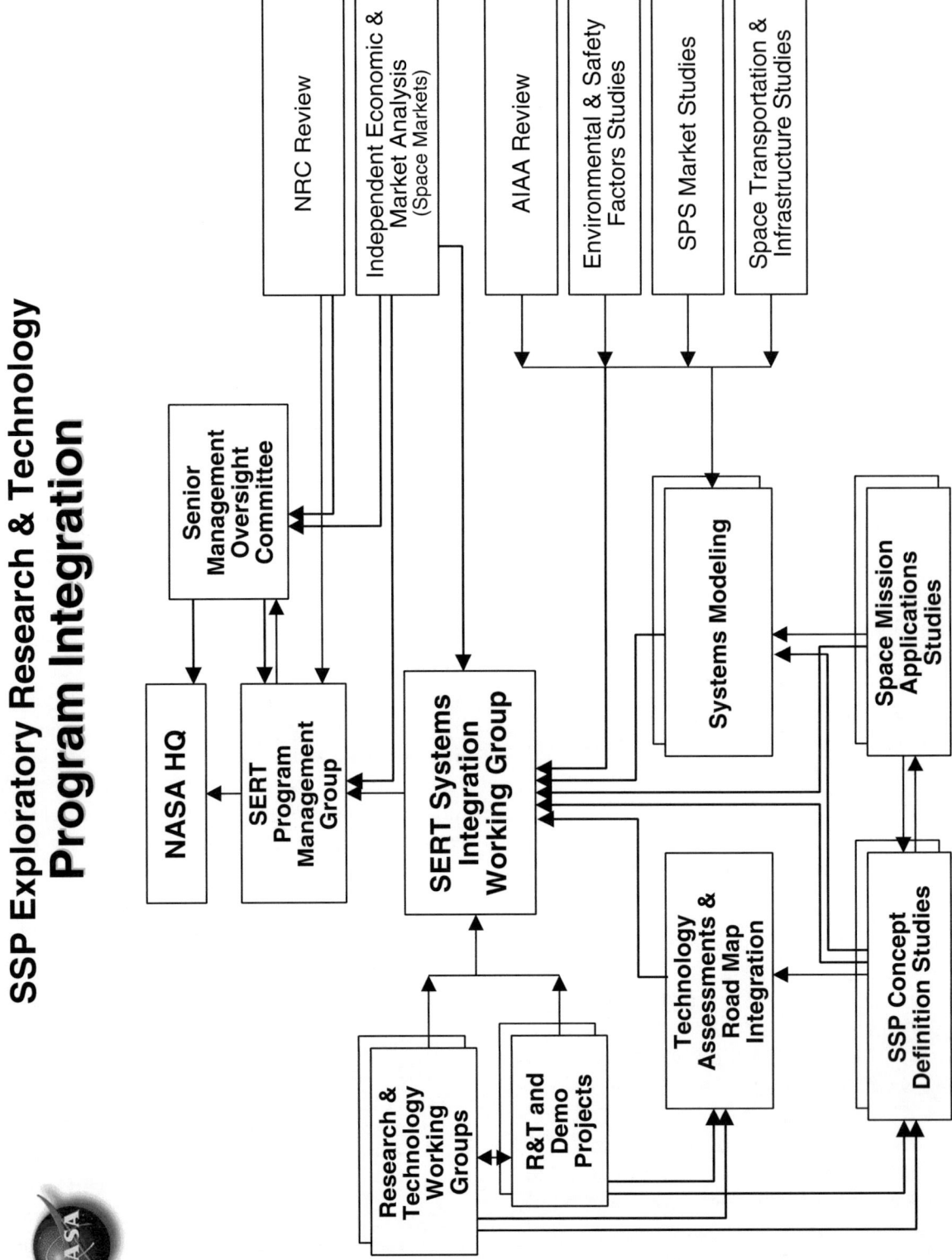

FIGURE 3-1 NASA's SERT program integration process. NOTE: Figure reprinted in original form. SOURCE: Mankins and Howell, 2000a.

in order for the currently planned program to become viable; (2) interaction with other government agencies, commercial entities (including the utility industry), and other organizations should increase; and (3) tough downselect decisions for various technologies and flight test hardware must be made, assuming that the increased funds will still be constraining.

However, when NASA leaves the planning stage and undertakes serious technology development, the SSP program would be better served by adding a technology development process complete with specific goals, dates, and procedures to be followed. This clarification of the organization and decision-making approach is necessary to meet technology goals at critical program milestones and with efficient use of funds. The program should define each of the integration functions and clearly charge the responsible organizations with explicit responsibilities. Although advancements in research and technology are difficult to schedule, adequate funding should be provided to each organization to undertake these assigned responsibilities at a high level of quality and at a reasonable pace. The approach should be similar to major program organizations at a much later stage of the development cycle, because the program involves several NASA centers, outside agencies, and the international community. NASA should develop a written implementation plan for carrying out the work.

Definition of a consistent process to adjudicate competing objectives is also necessary. It is almost inevitable that institutional considerations will sometimes be at odds with purely end-product goals, and it is better to address this likelihood with an objective decision process before positions stiffen within the SSP program and among the NASA centers.

Recommendation 3-1-1: NASA's SSP program should improve its organizational and decision-making approach by drawing up a written technology development plan with specific goals, dates, and procedures for carrying out technology advancement, systems integration, and flight demonstration. The SSP program should also establish a consistent process to adjudicate competing objectives within the program and specifically include timing and achievement of technology advances in robotics and space transportation in the roadmaps.

Sharpening the Technology Development Focus

Many SSP issues relate to manufacturing cost, not just performance. Cost estimation expertise is largely to be found in industry, not in the NASA laboratories or academia. Appreciable funding of and consultation with relevant industrial firms is required to gain authoritative costs to be used as future inputs to the integration model. Investment should also be made in high-payoff, high-risk manufacturing technologies to reduce the cost of all the components since many hundreds of duplicate pieces will be necessary.

The committee notes that technology is available today to build an SSP system (i.e., solar electricity is currently generated in space, microwave power transmission has been successfully demonstrated terrestrially, and space transmission is in demonstration by Japan). However, such technology would be impractical and uneconomical for the generation of terrestrial baseload power due to the high cost and mass of the components and construction. The system would also have to be constructed and utilized by humans in LEO, not the GEO planned for future large SSP systems, which would be constructed or deployed autonomously. It would be completely unable to "close the business case" for its (presumably) commercial sponsors and their funding sources. The committee considers that the present SERT cost and performance targets for several of the technologies are beyond present credibility, particularly in solar power generation (SPG), space power management and distribution (SPMAD), and wireless power transmission (WPT). In addition, goals in structural mass and performance may be too stringent. The committee suggests further design studies in this area before additional technology investment is made. Substantial improvements in assembly, maintenance, and service are also essential for the success of this program. If successful, these technology advancements will have many other applications. However, the committee realizes that the SSP program as currently envisioned is planned to unfold in stages, with NASA leading demonstration projects at progressively higher power levels (100 kW, 1 MW, and 10 MW) provided that positive decisions are made at each milestone to make the investments necessary to continue with the program.

Recommendation 3-1-2: The SSP program should allocate some of the technology development funds to the aerospace system level and component industries that can address manufacturing, assembly, and maintenance design and cost. This would include architecture, engineering, and electrical utility construction firms. An added benefit of broadening the industrial community involved in SSP development may be significant improvements to future baseline designs.

The committee observes that the earliest of the commercial satellites, should the private sector decide to join the SSP program, will be a subscale power station, perhaps a repeat of the 10-MW NASA experiment with improved technology, followed by a private-sector decision to construct and operate a 100-MW pilot plant. Only then will deployment decisions be made for power plants of 1 GW or larger. Each of these several decisions will be based on economic as well as technological figures of merit appropriate to their particular time—not based on the limited understandings of today.

Verification of SSP technology and the integration and testing of hardware and software are necessary before deployment of any SSP system. The SSP program's investment in developing spacecraft integration and testing methodologies and processes should be expanded. The goal should be to make verification of the performance of the SSP satellite possible with a minimum of ground testing. A combination of two approaches may be needed to accomplish this objective: (1) new modeling methods that accurately predict on-orbit function and performance and (2) a new design method to adaptively accommodate errors in the predicted function and performance.

The modeling approach may require the development of (1) high-resolution (possibly many millions of degrees of freedom) structural models (including the ability to predict damping in a zero-g environment), (2) complex, integrated structural-thermal-control performance models that include probabilistic sensitivity analyses, and (3) validated component-to-system verification test procedures. This effort should be supported by a complementary set of ground and flight experiment activities, the goal of which would be to establish the limits of the new modeling capabilities in predicting the zero-g performance of the SSP mechanical systems.

The adaptive approach may require the development of self-healing structures, self-tuning adaptive control systems, autonomous robots, and failure-tolerant structural concepts. Some of these developments may be very far into the future, requiring rigorous, long-term investment before they will yield practical success.

Recommendation 3-1-3: The SSP program should increase investments in developing spacecraft integration and testing so that the performance of SSP satellites can be verified with a minimum of ground or in-space testing. This may include the development of specialized integration, test, and verification methodologies for SSP spacecraft.

Because there are so many competing designs for various technology flight demonstrations (TFDs), it was difficult for the committee to determine the accuracy of the SSP system and cost models used to evaluate trade studies for various technologies. The current models should incorporate one detailed concept definition, making it possible to evaluate the payoff of specific technology efforts within the broad functional systems areas of SPG and WPT, among others. This concept definition should also include assembly, checkout, and maintenance techniques. The model needs to clearly show where SSP technology investments branch off to the benefit of other missions, including both in-space and terrestrial applications.

There was concern on the part of the study committee over the actual values used as technology metrics in the system and cost models. After discussion with NASA subsystem team leaders and systems integration personnel, the committee found the maximum and minimum values used seem to be especially suspect in areas such as structures and photovoltaics. Some performance metrics seem to be extrapolations that cannot be traced to the structural requirements specific to SSP platforms. Also, during (and after) presentations to the NRC committee, there were self-acknowledging errors in the models used to place priority on technical investments. This led the committee to be skeptical about the accuracy of the model input and the use of these metrics. There was further concern about the lack of comparison between assembly methods (i.e., fully autonomous, tele-robotic, or human) within the models. Technology metric input should be gathered from hardware builders, industry, academia, and other government agencies to improve the modeling effort. Input should include information on technical performance, current state-of-the-art technology, forecasts for technology advancement, and system cost modeling. The expansion of the modeling should also include other concepts (in addition to terrestrial power supply) that will benefit from the SSP technology investments.

Future comparisons in the models should include transportation costs and delivered performance on orbit, which will enhance decision making between alternative technology candidates. This seems to have been intended in some of the past SERT program trade studies; however, the committee believes that the principle is not sufficiently ingrained in the modeling effort.

Recommendation 3-1-4: The SSP program should review its technology and modeling assumptions, subject them to peer review, and modify where indicated. A single SSP concept should be rigorously modeled, incorporating technology readiness levels and involving industry in conceptual design, as a means to improve the credibility of the model input and output but *not* to prematurely select a single system for ultimate implementation.

Capitalizing on Other Work

The committee realizes that a full-scale SSP system is greatly dependent on future space transportation technologies not yet defined. Such technologies include ground launch to low Earth orbit (LEO) followed by transportation from LEO to geosynchronous Earth orbit (GEO). NASA's SSP program should identify and initiate detailed studies of the SSP program's space transportation needs and requirements. This NASA study should include the space transportation needs and requirements of the SSP demonstrator elements in addition to the space transportation needs and requirements of related space systems that NASA may later embrace for other uses based on SSP technologies. Industry involvement is imperative in defining the means and estimating the programmatics of new and evolving space transportation systems to fulfill these needs.

NASA's technology development program, called the Space Launch Initiative (SLI), is currently charged with addressing the Earth-to-LEO launch issue. Coordination, however, is apparently not yet present between the SERT program and SLI. Any future SSP effort should work closely with SLI as future milestones and roadmaps are developed for both programs in order to provide realistic goals for future implementation. A similar program is also necessary to develop space transportation from LEO to GEO. NASA defines work in this area as in-space transportation and infrastructure, which, if undertaken, may be separate from NASA's SLI initiative. These issues are discussed in further detail in Section 3.9.

The SSP technology program presented to this committee (Mankins, 2000) relies on parallel technology development efforts undertaken in several NASA enterprises and has strong interagency and, potentially, international implications. It will be important to the SSP program, therefore, to have a strong advisory infrastructure to help oversee the technology development progress efficiently. Currently, the main advisory body for the SERT program is the Senior Management Oversight Committee (SMOC), consisting of members from NASA and industry. Various other research and technical working groups and a systems integration working group are also used throughout the program integration process to infuse expert knowledge into the program (see Figure 3-1).

Upon evaluating the membership of the SMOC and, more important, the participation level of the individuals, the committee concluded that industry and government agency representation may be less than adequate to provide competent oversight and technical input. More external membership in these assessment teams and working groups should be brought together from industry and government in order to validate various technological metric inputs to the performance and cost models and to review progress at appropriate program milestones. Inputs from the Electric Power Research Institute (EPRI) and electric utilities will be of greater importance as the program proceeds. Government agencies, such as the Department of Defense (DOD), the U.S. Air Force (USAF), and the Department of Energy (DOE), should be involved, not just as liaison representatives but perhaps even as members of the SSP research team via the Intergovernmental Personnel Act or other cooperative programs. Such exchanges would not only foster cooperative activity among the agencies but also leverage all government investments in technologies applicable to SSP. It is also desirable, due to the breadth of SSP-related technologies, to establish advisory committees for specific technologies as well as the overall program. These advisory committees would be in addition to the research and technology working groups already used by the program.

This addition of non-NASA experts in the various technical areas will assist in providing technical peer review of the SSP program's research. Participation of others in the establishment of modeling metrics will also be extremely helpful to systems-modeling researchers by providing realistic and accurate technology parameters. As remarked on earlier, this is a major concern of the committee in areas such as solar arrays, SPMAD, WPT, structures, and robotics. Active representation of various aerospace manufacturing industries and the electric utilities will be imperative, especially as flight demonstration and implementation are approached.

Recommendation 3-1-5: The current SSP advisory structure should be strengthened with industry (including EPRI and electric utility) representatives plus experts from other government agencies (particularly DOD/Air Force, DOE, and NRO) in order to validate technological and economic inputs into the performance and cost models. Also, due to the wide breadth of technologies related to SSP, the program should establish similar advisory committees for specific technologies in addition to the research and technology working groups currently utilized by the program.

Various modeling techniques and methodologies have been previously developed for use in large space system integration and in cost, performance, and economic modeling. Such techniques do not seem to be used within the current SERT program or at least were not evident in the information provided to the committee. The SSP program should review previous modeling efforts and integrate successful methods, as appropriate. For example, it does not appear to the committee that the current SERT program has looked for lessons learned from modeling efforts used during the 1970s space solar power efforts (Hazelrigg, 1977) or considered other recent efforts in long-term, multi-phase technology development (Hazelrigg and Greenberg, 1991; Hazelrigg, 1992). These methods are *not* endorsed by the committee but are mentioned only to provide examples of other modeling on which NASA might capitalize.

The following sections address each of the major technical areas targeted in the organization of the SERT program. The discussion and recommendations include information on each technical area's objective and scope as part of an SSP system, the current technical state of the art, necessary goals for economic competitiveness, challenges to be met by the NASA team, and recommended priorities for NASA's SSP program activities.

3-2 SOLAR POWER GENERATION

Objectives and Scope

The solar power generation (SPG) subsystem of SSP consists of a large constellation of very large space solar photovoltaic (PV) arrays that convert the Sun's solar energy into electricity. The PV arrays would be connected to transmitters that would beam down as much as several gigawatts of power to Earth via microwave or laser transmission. The PV array consists of a large number of PV panels that each mechanically support hundreds of individual PV cells. The panels are electrically and mechanically interfaced together via cabling and a rigid support structure to form the large PV array. A constellation of these larger arrays would form the SSP solar power generation system.

Important figures of merit at the solar array level are specific power (watts per kilogram), cost (dollars per watt), areal power density (watts per square meter), and stowage volume (watts per cubic meter). For SSP, all four of these parameters are important, with complex trade-offs. For example, it can be argued that using increasingly higher-efficiency crystalline solar cells results in increased solar array cost in dollars per watt, but also gives increased specific power, areal power density, and stowage volume. On the other hand, if the paramount driver is PV array cost, as is the case in terrestrial PV applications, then lower-efficiency technology such as noncrystalline thin-film PV can provide advantages. At the PV blanket and cell level, the lower-efficiency (~10 percent versus 30 percent for crystalline cells) thin-film PV blanket can be 50-100 times cheaper than the crystalline cells. For relatively low power levels (1-20 kW), resulting thin-film PV arrays are expected to be at least 10 times cheaper, be 2-3 times lighter (2-3 times greater specific power), and have 3 times more stowage volume density than crystalline cell arrays, planar or concentrator, that produce the same power. However, an important disadvantage of thin-film PV technology will be much larger array size, 2-3 times larger than crystalline arrays, owing to the reduced cell efficiency. This in turn will increase array guidance, navigation, and control needs; structural mass and complexity; and cable mass, especially for SSP. Because the SSP system will be very large (several square kilometers), a significant amount of cabling would be needed, tens if not hundreds of kilometers.

To date, there has been no comprehensive PV array trade study for SSP. The baseline conversion technology indicated by the SERT program consists of array modules using high-efficiency crystalline solar cells under 8 times concentration. However, no studies were conducted to determine the support structure and electrical cabling that would be needed. Also, no consideration was given to the added support structure and guidance, navigation, and control system that would be needed to enable the Sun-pointing requirement of approximately 2°. Additional support structure would be needed to maintain flatness for concentrator arrays to maintain pointing tolerances, compared with planar unconcentrated thin-film arrays that suffer only a cosine loss.

Current State of the Art

Today's state-of-practice solar arrays provide specific powers of 30-60 W/kg and utilize single-crystal, 14-27 percent efficient (under the space spectrum) silicon, gallium arsenide, or $GaInP_2$/GaAs/Ge solar cells supported by rigid aluminum honeycomb or flexible polymetric substrates (Sovie, 2001; Oman, 2001). The cost for these arrays ranges from approximately $300 to $1,000/W (Marvin, 2001). Examples of state-of-practice arrays are the Boeing 601 10-kW array at approximately 50 W/kg and the International Space Station 260-kW array at 30 W/kg. Today's state-of-the-art solar arrays utilize either 26.5 percent (and soon 27-28 percent) efficient three-junction $GaInP_2$/GaAs/Ge solar cells supported by rigid aluminum honeycomb substrates or 17 percent high-efficiency silicon solar cells supported by a lightweight suspended mesh structure, resulting in specific powers of 70-100 W/kg and $500-$1,000/W (Sovie, 1999). Examples include the Boeing 702 15-kW array at 80 W/kg using three-junction cells and the ABLE Engineering Ultraflex Array at 100 W/kg using high-efficiency silicon cells. The solar array design baselined by the SERT program for SSP is Entech's stretched lens array (SLA) PV concentrator utilizing 28 percent efficient, three-junction solar cells under 8 times concentration via a silicon-based stretched Fresnel lens (Marvin, 2001). Measured performance for SLA at the module level is 378 W/kg but does not take into account the support structure mass required to assemble the modules into a practical array. Projected performance for a <20-kW SLA is less than 200 W/kg (O'Neill and Piszczor, 2001). However, most important, this projection does not include the additional structure mass required to maintain the 2° pointing accuracy over the several square kilometer array area of SSP. The specific power for the only full SLA solar array flown in space, aboard the Deep Space 1 mission, was 48 W/kg (Murphy and Allen, 1997).

The projected specific power for next-generation, flexible thin-film PVs at the blanket level is approximately 2,700 W/kg for 9 percent efficient, three-junction amorphous silicon on 1-mil Kapton and 770 W/kg on 0.5-mil stainless steel (Guha et al., 1999). The high specific power of thin-film PVs derives from the deposition of ultrathin (<10 μm) layers of semiconductor absorber material on thin (0.5-3 mil) flexible polymer or steel substrate blankets. The semiconductor layers are typically deposited onto large polymer or steel sheets using low-cost evaporation, sputtering, or plasma-enhanced techniques. The leading thin-film candidates under development are amorphous silicon (a-Si) and polycrystalline copper indium gallium diselenide (CIGS) and its alloys. State-of-the-art efficiencies for large-area (0.5-1.0 ft^2) a-Si are approximately 8-9 percent (space spectrum) and 6-8 percent (0.2 ft^2) for CIGS on stainless steel (Reinhardt, 2001a). The present development goal for space-qualified thin-film PVs under ongoing Air Force research and development programs is 10-12 percent by 2002, 15 percent by 2005-2007, and 20 percent multijunction CIGS by 2010-2015, yielding projected solar-array-specific power levels of approximately 300, 450, and 600 W/kg for an array support structure areal density of 0.1 kg/m^2.

Technical Performance Goals Needed for Economic Competitiveness

Successful development of an economically viable SSP will require substantial leaps in development of space solar array, PMAD, thermal control, wireless transmission, and launch technologies. However, improvements in PV solar array technologies alone will not enable SSP to be economically competitive with terrestrial utility electricity. The theoretical maximum solar cell conversion efficiency is between 50 and 60 percent for crystalline multijunction solar cells (Kurtz et al., 1997) and between 30 and 40 percent for amorphous and polycrystalline thin-film photovoltaics (Reinhardt, 2001a). Even for the case of 60 percent efficient crystalline solar cells, the array specific power will be limited to values less than 500 W/kg. Assuming that the SSP solar array must produce 3 GW of power (to result in 1.2 GW to the ground), the mass of a 500-W/kg solar array alone will be 6×10^6 kg, approximately 241 times that of the space shuttle's maximum cargo capability—clearly a formidable challenge. Even assuming the SERT program's launch-to-GEO goal of $800/kg, the cost of launching the array alone would be $4.8 billion. In the case of thin-film PV, using even 40 percent efficient arrays at 1,200 W/kg would require approximately 100 launches of the current space shuttle (at maximum payload capacity) for the array to reach LEO. It is clear from this simple analysis that improvements in PV-based power generation technologies alone, even to theoretical efficiency limits, will not enable SSP to be economically viable for competitive baseload terrestrial electric power, regardless of solar array cost. Even if the solar array were free, the overriding factor is the cost of placing it in orbit.

Challenges to Be Met

As stated above, improvements in PV solar array technology alone will not enable SSP to be economically competitive with terrestrial utility electricity, even if the solar array were free and theoretical efficiency and mass performance levels were obtained. The committee believes that the greatest challenge for the SSP program is to develop more realistic and accurate system cost and performance models, including theoretical solar array, power management and distribution (PMAD), thermal control, and wireless transmission cost and performance parameters, that will allow the launch cost to be realistically quantified. The issue is not the future cost of PV solar array technology, because one day the terrestrial PV industry will reduce costs to a point competitive with utility electricity, but the cost to place the array in orbit. Considering the paramount challenge in technology development required for other SSP disciplines, and the approximate $200-$300 million invested annually in space and terrestrial PV development worldwide, the SSP program should make minimal investment in current PV technologies. It is important to note that there is actually little difference between space and terrestrial PV technologies. There is virtually no difference in the electrically active part of the PV cell that controls conversion efficiency, called the p/n junction. The only difference lies in packaging of the cell, a technology that requires no significant new development.

Also, a more thorough trade study must be conducted to rule solar dynamic heat engines for power generation in or out. The option of using solar dynamic heat engines in the SERT program was briefly discussed; however, a comprehensive trade study has not been conducted. Solar dynamic options are presently 3-4 times heavier than conventional PV arrays, 20 W/kg versus 60-80 W/kg, respectively. Also, current solar dynamic options are roughly 10 times more expensive than PV arrays, $5,000/W versus $500/W, respectively. These data, along with the additional concern that solar dynamic requires very high solar concentration ratios and, hence, extremely accurate pointing of very large solar collectors, indicate that it may not be a good choice for SSP.

Recommended Priority for Investment

Considering the paramount challenge of reducing SSP power generation system mass to several orders of magnitude less than today's PV solar array systems, concepts other than conventional crystalline PV, thin-film PV, or solar dynamic concepts need to be developed. Even if SSP PV-based solar array were free, the theoretical specific power for conventional PV arrays of 1,000 W/kg would result in a cost-prohibitive launch requirement. Revolutionary breakthroughs are required in solar-to-electric power generation technology offering system-specific power in the range of 2,500-10,000 W/kg. Considering the small SSP investment in PV solar array technology today (<$2 million) compared with the large national and international PV technology investment of more than $200-$300 million annually, the committee recommends that the SSP effort focus future investments on revolutionary high-risk, high-payoff solar-to-electric approaches that could result in system specific powers in excess of 2,500 W/kg. At present, SERT is not working on such a strategy. The U.S. government's Small Business Innovation Research program could be one efficient avenue for exploring such high-payoff, high-risk technologies.

Recommendation 3-2-1: The SSP program should focus future investment in solar power generation solely on next-generation, revolutionary, high payoff, high-risk concepts.

Synergy with Other Programs

The current SSP baseline power generation system is PV solar. PV is considered a pervasive spacecraft technology and important for all near- and far-term DOD, NASA, and commercial space programs, as well as for terrestrial electrical power applications. All performance improvements in solar array technology, such as increased efficiency and reduced mass and cost, will greatly benefit a multitude of future government and commercial space and terrestrial power systems. There is presently significant U.S. government and private industry investment in the development of technology for space and terrestrial photovoltaic cells, arrays, and systems. The NASA and Air Force R&D annual budgets for space PV are approximately $2 million and $5 million, respectively, with industry investing approximately $1 million of independent R&D in space PV. The DOE has an annual budget of approximately $75 million for PV technology development, supported by approximately $20 million from private industry. Japan and the European Union have also increased support for PV research significantly during the past few years.

As stated in the previous section, the recommended priority for investment by the SSP program in power generation should be in revolutionary, extremely high-risk, extremely high-payoff solar-to-electric approaches that could result in system-specific powers in excess of 2,500 W/kg. Such investment would have tremendous synergy with future NASA and DOD SPG technology development. Both agencies have identified requirements for significantly higher levels of space platform electrical power for future missions, where the paramount goal is to provide the spacecraft payload increasingly greater power and mass budgets without increasing total spacecraft mass or cost. Any investment in next-generation, revolutionary, solar-to-electric conversion technology that results in increased solar array-specific power will benefit next-generation NASA and DOD spacecraft. Also, the SSP program should better capitalize on government and industry PV technology investment and participate in the U.S. government's Space Technology Alliance[1] studies that address government strategic planning and investment in space PV (Sovie, 1999). Further coordination with the National Center for Photovoltaics, which includes the DOE's National Renewable Energy Laboratory and Sandia National Laboratories, should also be pursued in order to capitalize on other PV technology investments.

Recommendation 3-2-2 [also 2-1-8]: NASA should expand its current cooperation with other solar power generation research and technology efforts by developing closer working relationships with the U.S. Air Force photovoltaics program, the National Center for Photovoltaics, industry, and the Space Technology Alliance's Space Power subcommittee.

3-3 SPACE POWER MANAGEMENT AND DISTRIBUTION

Objectives and Scope

The scope of the space power management and distribution (SPMAD) subsystem starts with the power source (solar panels or heat engine) and ends with the major loads—the wireless power transmitters and, if used, the electric propulsion system.

Most of this section is written assuming that the power source is photovoltaics. A short subsection near the end provides a discussion of the differences necessary in PMAD design if the power source is a heat engine using a thermodynamic cycle such as the Brayton (gas turbine) or Stirling.

The basic starting point is the individual solar photovoltaic cell (also known as a solar cell). The solar cell can be large (1 × 3 in. for crystalline solar cells and up to 12 × 12 in. for thin-film solar cells), and many must be connected in series to establish the desired bus voltage. A collection of solar cells, their substrate, and the supporting structure is called a "panel." A collection of panels with additional structure is called an "array." We assume that the PMAD interface is at the *panel* level of integration.

An essential element of PMAD is the cabling that collects the electricity and routes it to the major loads. Since the SSP system will be very large (several square kilometers) a large amount of cabling, tens if not hundreds of kilometers, is necessary and will account for a significant fraction of the mass of the system.

The major load of this subsystem is, of course, the wireless power transmitters that beam the power to Earth (or to a receiving station in space). For the mi-

---

[1] Currently the Space Technology Alliance consists of members from DOE, NASA, DOD, the Ballistic Missile Defense Organization, and the Defense Advanced Research Projects Agency.

crowave systems (magnetron, klystron, or solid-state power amplifiers), the transmitters must be physically close together, and all the solar panel power must be concentrated at one point. For the laser power transmission option, the transmitters will be distributed over many smaller satellites, thus allowing the length of the power distribution lines to be much shorter. A secondary load might be an electric propulsion system used for station keeping, obstacle avoidance, and attitude control.

Other major components of the SPMAD system are used for power conversion and conditioning. These components raise the voltage so that the power can be distributed with lower losses or less mass, or both, providing the exact voltages to the various loads. For some designs, the voltage may be converted from direct current (DC) to alternating current (AC) and then back to DC.

Current State of the Art

Most spacecraft have PMAD systems and have had them since the beginning of the space program. Thus, in principle, one could be built for the SSP system. However, it would be impractical due to excessive cost and mass. Both increased performance (higher efficiency, higher voltages, and higher temperature operation to reduce thermal control challenges) and reduced mass and cost of the individual components are necessary for a practical SSP PMAD system.

The design and performance of state-of-the-art PMAD subsystems varies widely, depending on the specific application. Many satellites have extensive battery systems, which require battery charging and discharge circuitry and controls. Efficiencies are 60-90 percent (Reinhardt, 2001b). For DC-to-DC inverters, the efficiency can be greater than 90 percent (Ashley, 2001). Masses of complete subsystems with cabling and battery chargers are 33-40 kg/kW (25-30 W/kg) (Reinhardt, 2001b). Cabling must be estimated and optimized separately for SSP because of the extreme lengths of the cable runs. Without including cabling and battery circuitry, the power density might be as high as 150-300 W/kg (Ashley, 2001). Costs are $50-$100/W (Reinhardt, 2001b; Ashley, 2001).

Technical Performance Goals Needed for Economic Competitiveness

The NASA subsystem team has adopted the following goals for this SPMAD subsystem: end-to-end efficiency greater than 94 percent, mass less than 2 kg/kW (power density greater than 500 W/kg), and a cost of less than $300/kW (30 cents/W) (Mankins and Howell, 2000b).

Challenges to Be Met

System Design Optimization

The mass and cost of the SPMAD subsystem form a major portion of the whole SSP system. For example, at 500 W/kg the PMAD mass would equal that of the solar array. If the solar array achieved a breakthrough performance goal of greater than $2,500/W, the SPMAD system would dominate the mass of the array.

As a result, optimization of the *design* of the system will be important. This is a systems design issue and must have continuing attention to help guide investment priorities for the various hardware components of the whole SSP system. In other words, this optimization is not an R&D activity but rather a systems activity. Adequate resources are necessary to do it well, but it should not have to be an area of significant investment.

Power Distribution Voltage and Insulation

Power distribution voltages and insulation are areas of future research. The conductors will undoubtedly be aluminum, which is optimum. High voltages (tens of kilovolts) are desired to reduce transmission line losses and mass. Some design concepts even use bare wires (i.e., no insulation). Use of these high voltages will present challenges of electrical breakdown and performance of insulators, insulation, and switchgear.

The power collection and distribution voltages are important system optimization variables. R&D investments in this area include (1) space effects on conductors (such as arcing and damage on bare wires from the plasma and particles resulting from possible meteoroid impact to the SSP system as a whole) and (2) types of insulation for insulated systems. These topics are important because they may limit the highest voltage, having a first-order impact on the system mass and cost. Electrical arcing and other breakdown limits are also key issues. Steady-state leakage of current can occur without arcing. Other solar system and spacecraft events such as solar storms, thruster firing, outgassing, and possible cosmic ray or meteoroid impacts can cause potential damage. Recently, communications satellites in GEO have experienced serious arcing damage (Hoeber et al., 1998). As a result, select flight experiments at GEO may be necessary to ascertain optimum allowable voltage limits.

Reduction of thermal heat rejection and radiation hardness of components are major challenges in PMAD component design. High-temperature operation and passive thermal control are goals. Adequate investment must be made in these areas to find the appropriate solutions. In-space experiments may need to be done, at GEO conditions, to demonstrate lack of breakdown and possibly lifetime issues.

Power Conversion and Conditioning

Power conversion and conditioning are important parts of the PMAD subsystem. These aspects of PMAD are major mass and cost drivers. The components needed include power transistors (probably thyristors), sensors, transformers, inductors, capacitors, and switch gear (transistors and circuit breakers). These components can be assembled into inverters to convert DC to AC (and vice versa) and to raise and lower the voltages. They also filter and regulate the output voltages to those required by the loads.

These components have been an area for investment since electricity was discovered—initially for terrestrial power applications and, for the last 50 years, missiles and space. Total PMAD system efficiencies remain below 90 percent. Enormous investments have been made over many years by both the government and the private sector. Steady progress has been made; however, this is an appropriate area for investment by the SSP program in order to reach its goal of greater than 94 percent efficiency.

A few percentage points of improvement in PMAD efficiency will translate to a larger percentage point reduction in SSP mass and cost, reduced system heat load, and thus fewer thermal management problems. However, large breakthroughs should not be expected but, rather, slow, incremental progress. Efficiencies may improve a few points, but factor-of-two improvements will not be seen. Thus, reductions in mass for this subsystem, other than by optimization, will be modest.

This, however, is an area where investment in *manufacturing* techniques may provide dramatic reductions in cost. Space components, typically, are made in very small quantities and are subject to extensive testing. For any SSP system, there will be a need for millions of each item, so process improvement and mass production could result in substantial cost reduction. NASA typically does not invest in manufacturing technology, but for SPMAD, that investment may represent the largest payoff.

Each subsystem should have mass, cost, and performance targets. The mass is also important because a significant portion of the overall system cost, approximately one-half, is incurred in launching all the mass to GEO.

Heat Engine Usage

Some of the SSP conceptual designs being considered use a heat engine, such as a Brayton or Stirling thermodynamic cycle, to convert solar heat to electricity. An advantage of such a system is that the sunlight can be concentrated at one, or several, points, and the power conversion and distribution (cabling) mass and costs can be reduced. Disadvantages are the large and heavy heat rejection radiators required, lifetime issues with moving parts, and the management of angular momentum of the rotating machinery for the Brayton option (the Stirling will undoubtedly have a reciprocating alternator).

A system-level advantage of using a dynamic system is that the electrical generating system will directly produce AC power at a relatively high voltage. This will eliminate the inverter that is needed in the DC systems to convert DC to AC.

Recommended Priority for Investment

Recommendation 3-3-1: A major investment should be made in space power management and distribution. This should include efficiency improvements, mass reduction, and manufacturing techniques designed to reduce the cost of the components. Almost all of the manufacturing investment should go to companies that are already making such components for either ground or space applications.

Synergy with Other Programs

SPMAD is important for all space programs—whether they be NASA, DOD, or commercial efforts. Thus, investments made in this SSP program will benefit other users. In addition, PMAD R&D by others will benefit the SSP program. It is important to have frequent coordination meetings with other SPMAD efforts.

3-4 WIRELESS POWER TRANSMISSION

Objectives and Scope

The wireless power transmission (WPT) subsystem starts with the electrical input to the transmitting devices in space (beamers), and ends with the DC output from the receiving elements on the ground (or in space for a space-to-space application).

The transmitting elements under consideration are of four types: (1) magnetrons, (2) klystrons, (3) solid-state amplifiers for microwave systems, and (4) lasers for the laser option.

Microwave Option

The microwave option may operate at 5.8 GHz (5.2-cm wavelength), although other frequencies are being considered. The microwave receiving elements, called "rectennas" (rectifying antennas), convert the transmitted microwave energy into electric power through use of a mesh of diodes and dipoles. They need to be arrayed at about 0.6-wavelength spacing (3.1 cm) in order to have high collection efficiency (Dickinson, 2000).

This subsystem also logically includes the intervening media—the space environment, the ionosphere, and the troposphere (lower atmosphere). The power beam may affect the media and the media may affect the beam, causing losses in power. The microwave systems will be nearly all-weather and continuously available except for short semiannual periods when there will be occultation of the Sun or when the beam must be interrupted for safety reasons.

For the microwave option, the individual transmitter element power level determines the transmitting array size and the amount of beam steering required. The elements must be contiguous to avoid grating effects causing large side-lobes (with resulting losses and electromagnetic interference concerns) at the receiving site. For the baseline system at 1.2-GW output to the utility interface, this results in a dense concentration of transmitting elements in a 500-m-diameter circle.

Laser Option

The laser option will probably operate in the near-infrared spectrum at 1.03 μm (Er:YAG laser) or 1.06 μm (Nd:YAG laser) wavelength.

The laser option has certain advantages in scaling but is expected to have an end-to-end power transmission efficiency about half that of microwave systems (20 percent versus 40 percent) and suffers from other disadvantages such as weather dependence (requiring redundant receiving stations) for terrestrial applications.

The laser option, however, has many advantages for space-to-space applications. The beam width can be narrowed by increasing the transmitting aperture (providing much better scaling to lower powers than the microwave option). The receiving elements for the laser options are solar cells. For the frequencies under consideration, silicon solar cells will most likely be chosen.

In current SSP designs developed during the SERT effort, the laser systems are incoherent with one another, and each beam has a very small divergence angle. In order to limit intensities to eye-safe levels, the SSP system must be broken into many (20 to hundreds) smaller satellites and flown in a "halo" orbit at GEO.

To prevent any chance of using the laser system as a weapon, the aperture of the individual laser elements will be kept small. This will make the beam spot of an individual laser satellite about the same size as the receiving array. All of the satellites will be focused on that same receiving array so, by design, the beam intensity cannot be increased above its design level.

The laser system is affected by weather—rain and moisture will cause reductions and sometimes complete interruption of power. NASA proposes to mitigate this problem by having multiple receiver fields at widely spaced locations to reduce the probability of simultaneous cloud cover over all the receiver fields.

Current State of the Art

WPT has had numerous microwave demonstrations over the past 35 years. The highest power transmitted was 34 kW at S-band (Dickinson, 2000). The laser option has been demonstrated, but at much lower power levels. The laser option is preferred for lower power levels because it scales better to smaller sizes. For that reason, it is preferred for space-to-space applications.

Technical Performance Goals Needed for Economic Competitiveness

For the microwave system, the project has defined an overall transmission efficiency allocation that will yield 40 percent efficiency end to end from PMAD DC to receiver DC. A mass goal of 5 kg/kW and a cost goal of $2,500/kW have been established (Dickinson, 2000). These are not consistent with a system-level cost goal of 5 cents/kW-hr. The NASA program should make consistent performance, cost, and mass allocations.

Challenges to Be Met

For the transmitting side, challenging issues include the efficiency of the transmitters, thermal control and heat rejection, the antenna array, and beam steering. The klystron and the laser system will probably require active cooling systems, such as pumped fluid loop radiators. It is hoped that the solid state and magnetron can be passively cooled with careful thermal design. Stringent mass and cost targets must be met.

The receiving elements are rectennas for the microwave systems. The receiving elements for the laser option will be silicon solar cells that also take advantage of the sunlight falling on them. Mass is not an issue for the ground subsystem, but cost is an issue.

A continuing activity will be needed to reserve the necessary frequency allocations for the early SSP demonstrations, as well as for the ultimate full-up system. There are also issues to be addressed relating to electromagnetic interference and electromagnetic compatibility effects on other ground and airborne systems.

Research will be necessary to develop intrusion detection systems to keep people and airplanes out of the high-intensity parts of the beam. If a plane is about to enter the beam, the current plan is to rapidly defocus the transmitted beam. This will minimize electrical, thermal, and attitude disturbances on the satellite. A major hurdle for the WPT subsystem is public acceptance. There are concerns about the safety and health effects for both the microwave and laser options. The project has established guidelines to limit the beam power densities to levels thought to be safe. The ground portion of the system must be worker-safe so that maintenance can be done while the beam is on. There are also concerns about low-level exposure "outside the fence." The SSP program should continue to use design specifications for the WPT subsystem that meet public health safety standards. Refer to Section 3-10 for further discussion of general environmental, health, and safety issues for SSP systems.

Recommended Priority for Investment

Recommendation 3-4-1: As long as the annual budgets remain modest (<$10 million), NASA should concentrate its resources on the best of the microwave options and the laser option. Both should be demonstrated, and the laser option should be brought to the same level of maturity as the microwave options. When the resources become more substantial, NASA should re-examine the other microwave options to make sure that the one selected is still optimal.

Recommendation 3-4-2: The SSP program should develop the laser option to the same level of maturity as the various microwave options. Investment should be made in increasing the laser conversion efficiency, improving associated heat rejection systems, and investigating possible uses of direct solar pumping.

Recommendation 3-4-3: Point-to-point power transmission should be demonstrated with the laser option. Initially this should be conducted on the ground. The need for a space-to-space demonstration should be studied, and such a demonstration should be conducted if warranted.

An advantage of the laser option is that a meaningful demonstration can be done at relatively small scales and at moderate cost. This will benefit both the space-to-space and ground-to-space applications.

Recommendation 3-4-4: Larger-scale, system-level wireless power transmission demonstrations should be planned, either ground or space-to-space demonstrations. Some might involve the International Space Station.

These demonstrations should be selected to have *commercial* potential wherever possible. This may attract cost sharing and possibly create an application that will serve to develop a larger market and help lower costs. *The space demonstrations are likely to be quite expensive.*

Synergy with Other Programs

There is interest in WPT within the DOD and various commercial ventures (Dickinson, 2000). Most applications have to do with powering an airship or airplane so that a platform can be kept above a given area for long periods—say, weeks or months. There are also some applications that involve powering a satellite either from the ground or from another satellite. If the right demonstrations and technology options are chosen, it is likely that other significantly funded projects could be established that would be synergistic with the NASA SSP program.

There is some interest in SSP and WPT technology outside the United States (Erb, 2000). There are continuing space WPT sounding rocket experiments by the Japanese. The French are planning to beam 10 kW of power across a bay (700-m range) at La Réunion island (Pignolet, 1999) See additional discussion of international interest in Section 2-3.

3-5 GROUND POWER MANAGEMENT AND DISTRIBUTION

Objectives and Scope

The ground PMAD subsystem begins with the output of the receiving elements of the WPT subsystem—either a "rectenna" (rectifying antenna) for the microwave options or solar cells for the laser option—and ends at the interface onto the utility power grid. In the utility and terrestrial PV communities, this collection of subsystems is usually referred to as balance-of-system. For space-to-space power transmission, the PMAD for the receiving end will be similar to the PMAD of conventional satellites.

In contrast to the space PMAD subsystem, the ground PMAD subsystem does not need a mass goal. However, it does need cost and efficiency goals.

Current State of the Art

The components needed for this subsystem are the same categories as for the SPMAD—namely, cabling, insulation, inverters, transformers, power transistors, inductors, capacitors, switch gear, sensors, and so on. For the microwave options, the receiving array will be larger than the solar arrays in space, so the cabling lengths will be even longer. The same issues of high voltage and AC collection will be present here as well. The final output could be high-voltage AC or DC depending on the utility network to which the system is being connected.

For the laser option technical needs are similar. The receiving panels will consist of solar cells and for all practical purposes will behave like terrestrial PV. The array will be much smaller than for the microwave option, so the cabling lengths can be shorter. The shorter cable lengths will undoubtedly result in lower optimum DC collection voltages.

The terrestrial PV industry is currently focusing on distributed applications of 10-100 kW. There no longer seems to be government or industry interest in large PV farms of hundreds or thousands of megawatts.

Technical Performance Goals Needed for Economic Competitiveness

The components in the ground PMAD subsystem are familiar to the utility industry for conventional as well as terrestrial PV applications. The utility industry has a desire for lower cost components, and competitive pressures by suppliers should result in R&D investments necessary to improve performance and reduce unit costs. This investment will probably wait until a larger market is perceived.

The National Renewable Energy Laboratory (NREL) estimates a cost of 15-25 cents/W for 1-MW and greater power systems (Koproski and McConnell, 2001). NREL has funded a PV Manufacturing Technology (PVMat) program in conjunction with Sandia National Laboratories, which included funding of inverter technology at approximately $1 million/yr for several years. Southern California Edison uses an estimating figure of 30 cents/kW-hr for the balance of the system. A consulting firm reports that the largest U.S. installation of PV is 1 MW at the University of California, Davis, with an estimated cost of 10 cents/W at high volumes (Whitaker, 2001).

The SERT program has established an efficiency goal for the ground PMAD subsystem of 85 percent and a cost goal of less than 0.5 cents/kW-hr. It appears that the cost goal is not stringent enough.

Challenges to Be Met

The ground PMAD subsystem will be similar to terrestrial solar power systems, so the challenges are similar. It will also utilize components that are familiar to the utility and supplier industries. The biggest challenge is cost.

Another issue is energy storage for the SSP system. The SSP program wants to apply the gold standard of baseload power to compete with other ground-based alternatives. This will require storage to provide continuous power during annual eclipses of the satellites (a few hours per day for a few days) and for occasional beam interruptions for safety reasons (e.g., an airplane about to fly through the beam). Therefore, NASA has included ground energy storage in its work breakdown structure.

When the first SSP system is operational, it will represent a tiny fraction of the nation's or a given utility's electricity supply. The utility will probably integrate it into its system *without* storage. As SSP becomes a larger fraction of the electricity supply, a given utility system may want to include storage, but probably *not* at each individual receiving station. Storage is a network and system issue. Using storage a long distance away from the SSP receiving station may well be optimum. The solution will most likely be utility specific. The utility industry also has other needs for energy storage and will undoubtedly make the investment necessary to develop and build such systems when they are needed.

Recommended Priority for Investment

Recommendation 3-5-1: NASA should neither invest in R&D for ground PMAD technologies nor spend resources on ground-based energy storage.

If, for comparison purposes, NASA wants to quote cost figures for baseload systems with storage, then storage should be included only *analytically* in the economic models. NASA should not conduct research and development in this field. The Terrestrial Photovoltaics Roadmap being followed by DOE covers this area of investment.

Synergy with Other Programs

The DOE NREL and Sandia National Laboratories jointly manage the National Center for Photovoltaics. This center funds various photovoltaic deployments of distributed PV (typically 10- to 30-kW installations). These deployments are no longer considered "demonstrations." Each application increases the market, and technical improvements occur. They also conduct the PVMat program on manufacturing technology, and about $80 million has been spent since 1992, including about 60 percent cost share by industry (Bower, 2001; Koproski and McConnell, 2001). Almost all of their work is directly applicable to the SSP ground PMAD subsystem.

The utility industry and the Electric Power Research Institute have changed their emphasis since the deregulation and restructuring of the electrical utility industry. They used to fund aggressive programs in renewable energy, including PV, but this is no longer the case.

3-6 SPACE ASSEMBLY, MAINTENANCE, AND SERVICING

Objectives and Scope

The technology discussed in this section is that required to assemble, maintain, and service a full-scale SSP system, with NASA's stated goal being to use robotics for all of these operations. Considerable R&D must be performed to bring robotic capabilities to a level consistent with this goal. These efforts can be grouped into the following general areas: robotic system architectures, robotic component technologies, and robotic control modes. It is not the objective of this section to identify which *specific* robotic architectures, technologies, or control modes warrant investment. This selection process should be performed in conjunction with systems studies that include the cost of robotics and in accordance with the technology downselect procedures described in Section 2-1 of this report (technology readiness level, research and development degree of difficulty, and the integrated technology and analysis methodology).

The design details of any SSP system should integrate assembly and maintenance requirements from the outset. Design of the robots will clearly be a function of the hardware to be assembled; however, the converse is true as well: SSP hardware will need to be designed to accommodate the type of robotic assembly baselined for the time of construction. For example, robotic assembly may require the installation of grapple hard-points at regular locations on the structure, and maintenance may require that robot transportation rails be incorporated. Clearly, it will be essential to consider robotic assembly and maintenance early in the design phase of space solar power system concepts. The cost and logistics related to assembly operations should also be included in the program's early systems studies.

The NASA SERT program's robotics goal for the gigawatt-size solar power system (MSC 4) is to perform the entire assembly using a fully autonomous robot team (Culbert et al., 2000). This is an ambitious goal, even in the 20-year time frame allotted. The current (i.e., December 2000) SERT program does not consider human assembly, maintenance, and servicing for its *full-scale* system (except the use of ground personnel). Human assembly is expected by the program for early flight test demonstrations. Because no analyses, systems studies, or data have been shown to this committee demonstrating that full autonomy is the least expensive or optimal assembly method, discussions in this section address the more general question of optimal assembly techniques for large SSP systems. Conceivably, SSP assembly could be accomplished using robotic capabilities anywhere in a range from completely tele-operated (at one end of the robotic spectrum) to fully autonomous (at the other end). The optimal choice of autonomy level, as well as the choice of where to locate humans controlling the robots (on the ground or in orbit) should be made using results of systems studies that explicitly include the cost of both human and robotic involvement.

According to the SERT roadmap for space assembly, inspection, and maintenance, the early flight demonstrations (MSC 1 and MSC 1.5) do not use robotic technology for assembly but rely instead on deployment and extravehicular activity. While this approach is understandable because robotic technology needs are postponed to a later date, it removes robotics from the invaluable operational experience developed with regular flight experiments. Robotics should be considered as a possible component of every SSP development mission and used, if applicable, for each flight demonstration, consistent with the technology available at the time and integrated into the space operations necessary for that mission. To achieve the level of robotic capability necessary for assembly of the full-scale system, other flight experiments, in addition to the MSC flights, will likely be necessary.

Current State of the Art

Operational space flight robotics to date have been limited to simple sampling systems (Surveyor, Viking) and a large crane-type positioning manipulator (the shuttle's Remote Manipulator System [RMS]) with an end effector specialized for a single grapple fixture design. The next-generation positioning manipulator will be the Space Station RMS, which will incorporate the ability to maneuver end over end but will still be limited to a single grapple fixture design. This system will eventually support the Special Purpose Dexterous Manipulator, which will be capable of many specialized servicing tasks.

Assembly of space structures by robots in a realistic simulation environment has been limited to the Beam Assembly Teleoperator and Ranger Neutral Buoyancy Vehicle of the University of Maryland (Spofford and Akin, 1984). The Ranger Tele-robotic Shuttle Experiment (TSX) is under development for a 2003 flight demonstration and represents the type of enhanced dexterous capability that could be incorporated on an early SSP flight demonstration.

Future applicable robotic systems that are currently in laboratory development are the Carnegie Mellon University Skyworker and the NASA Johnson Space Center Robonaut; the latter is particularly interesting in that it features near-human manipulative capabilities with anthropomorphic articulated five-fingered hands. Systems such as these could be advanced to flight demonstrations within a decade and should be available for assembly of SSP flight experiments.

Far more advanced robotic technologies and systems include fully autonomous humaniform robots (descendents of Robonaut), self-replicating systems, nanobots, and cellular automatons. While these are topics of active theoretical research interest, they require breakthroughs well beyond current technology and may not offer any significant advantages for SSP over more near-term robotic systems. More applicable perhaps to SSP assembly, maintenance, and servicing are novel concepts for self-assembling, self-monitoring, and self-repairing structures or electrical systems. Thus, for example, smart structures with embedded sensors or self-reconfiguring electronics that could reroute around damaged solar cells could reduce the need for robotic inspection and repair. Integrated "smarts" of this nature already exist in modern military aircraft and will probably become indispensable for spacecraft of the future, particularly large remote solar platforms.

Control of most robotic systems used to date in space has been through some form of tele-operation (human directly commanding manipulator motion). Low-level supervisory control (human commanding a well-defined task, then supervising task completion) is commonly used with ground-based manipulators, and high-level supervisory control (human commanding a goal and letting the robot figure out how to accomplish it) has been demonstrated in limited laboratory conditions. Full robotic autonomy requires some form of artificial intelligence in the robot. While much progress has been made, full autonomy has not yet been demonstrated for space robotic systems.

For all but the most autonomous systems, humans must be present somewhere and communicating with the robots in some manner to control them. Many forms of remote control have been demonstrated, from operating a manipulator in the next room to driving a rover on Mars. In general, once the operator is removed from the work site, the separation distance is important when the control time delay becomes noticeable. Control time delay will be an issue at GEO for tele-operated systems.

Technical Performance Goals Needed for SSP

Comprehensive robotic performance is difficult to measure because there are numerous aspects of importance, including mobility, manipulator capabilities, control modes, speed, strength, precision, and reach envelope, among others. One cannot define quantitative threshold values that must be met in any of these categories to enable SSP robotic construction because these values are entirely dependent on the hardware to be assembled. It is possible to discuss qualitatively, however, the types of robotic activities and levels of performance that may be necessary to enable SSP assembly and maintenance.

While a solar power satellite is orders of magnitude beyond anything ever built in space, it is similar in scale to a number of everyday Earth construction projects, including large bridges, deep-water drilling platforms, skyscrapers, or large ships. These engineering projects are equivalent in mass and similar in size to SSP and are composed of a much larger number and variety of components under much greater load. These structures are quite effectively built without huge teams of finely interacting robots, or self-replicating systems. Indeed, optimal design for these systems does not call for hundreds or even dozens of interacting agents working with precise coordination but instead uses small teams of agents (humans) working semi-autonomously at various locations around the structure. Assembly and maintenance of SSP should be modeled on these types of systems.

Robotic technologies have already been flight proven (RMS), dexterous and capable robotic systems are nearing flight readiness (Ranger Tele-robotic Shuttle Experiment), and highly capable anthropomorphic systems are currently in laboratory testing (Robonaut). All of these, as well as Earth-based simple technologies such as pick-and-place automation, have real and practical benefits in an SSP assembly scenario and should be fully evaluated as components in a robotic work site.

It is interesting to note that early SSP assembly studies performed in the 1970s, when the SSP concept first arose, envisioned robotic systems such as Ranger, Robonaut, and Skyworker as being necessary and sufficient for the SSP assembly process (Akin et al., 1983). Yet now, when these systems are nearing operational status, current assessments of robotic technologies necessary for SSP assembly indicate the need for large numbers of cooperating autonomous robots (Culbert et al., 2001), a capability still very far off today. There is no doubt that further engineering development can lead to advanced autonomous abilities, greater aptitude for cooperation among robots, and reduced demands on support (physical and mental) from Earth. All of these advancements will improve the capability to assemble large space systems effectively and efficiently. However, it is paramount that the technology investment today not be focused exclusively on the most ambitious goals. Certainly some funding might beneficially be invested in revolutionary breakthroughs in robotic technologies, but most investment should be to nurture and support the less esoteric robotic capabilities currently nearing flight status and the technologies necessary for the MSC flight demonstrations.

Some robotic technologies can be identified as being so basic and useful that they will most likely be developed regardless of how an SSP system is ultimately designed, assembled, and maintained. Such technologies include:

• Manipulator end effectors and dexterity to allow for delicate handling of lightweight SSP structures such as inflated tubes, film-like blankets (reflectors, thin-film solar cells, thermal blankets), alignment-sensitive mechanisms (concentrator arrays), and other difficult-to-handle components (stiffening cables or wire bundles);
• Robot-compatible connectors and interfaces, not just for the mechanical connection of components,

but also for electrical wiring, fluids routing, and thermal conduction;

• Improved vision systems, including image recognition, work site position and pose calibration, and surface inspection in extreme lighting conditions;

• Advanced control and human interfaces for enhancing robotic performance, including mitigation of time delays and accommodating limitations in real-time video feedback;

• Mobility systems, including positioning manipulators (e.g., RMS), "walking" systems (Skyworker), and free-flight mobility (AERCam, Ranger TSX);

• Decreased maintenance and servicing requirements (i.e., longer times between scheduled and unscheduled maintenance operations) and more robust actuator systems;

• Decreased power requirements (longer time between refueling or recharging);

• Basic research in remote work site technologies and use of robots in weightlessness, including human-robotic cooperation; and

• Ongoing development and performance evaluation of innovative robotic architectures and concepts, such as serpentine manipulators, modular reconfigurable systems, and free-flying miniature inspection robots.

Challenges to Be Met

One of the biggest impediments to robotic availability for SSP assembly is the lack of NASA support infrastructure for robotic technology development today. While there are identified agency programs (such as the NASA Institute for Advanced Concepts and the Office of Aerospace Technology) to support technology at technology readiness levels (TRLs) 1, 2, and 3,[2] established flight programs try to achieve mission assurance by minimizing the use of advanced technologies to the degree possible. This makes TRLs 4, 5, and 6 a no-man's land,[3] where promising technologies lie fallow for years for lack of advocacy or established programs to nurture them through to flight demonstration. It is essential that a well-supported and comprehensive program in robotics and automation technologies be re-established and aimed not just at enabling advanced research, but at taking basic research all the way from concept (TRLs 1 and 2) through flight demonstration (TRL 9).[4]

A viable SSP concept design will have to consider manufacturability: design from the outset to facilitate assembly and checkout on orbit. The design of the platform will be interwoven with the design of the assembly process. Trade studies (such as type and quantity of robots, specific capabilities of robots, and balance of roles and responsibilities between humans and robots) must be performed in the context of the overall system, allowing the trade space to encompass the design of the platform itself, as well as the components of the assembly system. Results of these systems studies will help identify what level of robotic capability is necessary and how much human involvement is optimal, thus indicating the robotic technologies in which to invest. No evidence of this type of analysis was presented to the committee, yet it is essential to establishing appropriate research goals for robotics.

Determining the optimal mix of humans and robots in a space assembly workforce is actually a complex study that would need to include not just the design parameters of the space platform but also the cost of having robots or humans, or both, in orbit with all their supplies and support equipment. Since the cost of transporting and maintaining humans in orbit is high, it is undoubtedly ineffective to use humans for simple and repetitive assembly tasks on orbit. However, in the absence of detailed systems studies, there is insufficient justification for insisting a priori that all tasks be performed exclusively by robots, particularly intricate and complex tasks such as repair and maintenance of robots and production equipment. It would require a great deal of development effort to reach this level of autonomous robotic capability and might not (arguably) produce a higher likelihood of assembly success than a system that provides the opportunity for humans to intervene directly when necessary. Experts in the field of automation and robotics for space applications

---

[2]Technology readiness level 1 (TRL 1) refers to basic research where basic principles have been observed and reported. TRL 3 commonly refers to technologies that have advanced through a characteristic proof-of-concept evaluation.

[3]TRLs 4-6 consist of technology development and demonstration both in the laboratory and in a relevant environment.

[4]TRL 9 involves system test, launch, and operations during which an actual system is flight proven through successful mission operations.

stress the value of having some direct human presence for real-time accommodation of unexpected events, robustness, and capacity for innovation, all of which will yield improved mission assurance (Akin and Howard, 2001).

Recommended Priority for Investment

Recommendation 3-6-1: The SSP program should perform systems studies to investigate the trade-off between robotic autonomy and human involvement for SSP assembly, maintenance, and servicing operations. It is essential to determine the optimal level of robotic autonomy for SSP operations and the optimal number and location of humans involved in these operations.

Recommendation 3-6-2: The SSP program should perform near-term flight demonstrations of robotic assembly techniques, as well as robotic maintenance and servicing operations. Robotics testing should be incorporated into all SSP flight demonstrations, if possible and as applicable.

Recommendation 3-6-3: SSP systems should be designed from the outset for robotic assembly and maintenance.

Synergy with Other Programs

Robotics applications are everywhere, and spin-off technologies from SSP will find uses in at least as many ground-based applications as space-based applications. Examples follow:

- Robotic maintenance and repair of the International Space Station (ISS) and its future derivatives;
- Robotic maintenance and repair of scientific spacecraft (such as the Hubble Space Telescope or Next Generation Space Telescope) and commercial spacecraft (communications satellites);
- Robotic exploration (and possible use) of other planets and space objects (comets, asteroids, libration points, missions out of the solar system, etc.);
- Robotic capabilities for undersea operations and exploration of other Earth environments dangerous or toxic to humans (e.g., Antarctica, volcanoes, nuclear power plants);
- Factory automation and ground-based manufacturing;
- Robotics for medical applications (precision surgery); and
- Robotics for military applications (ground combat troops).

Because these applications are so widespread and potentially powerful, a number of organizations and countries are already investing in robotics development, including NASA Johnson Space Center, the Jet Propulsion Laboratory, and Defense Advanced Research Projects Agency in the United States, as well as organizations in other countries, including Japan. Much of this investment may be applicable to SSP assembly and maintenance operations. However, a definite need remains for new investment in robotic technologies that are unique to SSP, including flight experiments with early SSP demonstrations, systems analyses to quantify costs and identify critical robotic technologies, and development of man-machine interfaces that allow progressively more autonomous systems.

3-7 STRUCTURES, MATERIALS, AND CONTROLS

This section reviews the investment strategy for the SSP structural concepts, materials, and controls research effort. Rather than providing a specific critique of individual program technologies, the intent of this section is to provide a basis for prioritizing technical investment within this area. Specific findings and recommendations are provided, with a focus on the unique requirements of SSP structures.

Historical Perspective and State of the Art

Much of the basic structural technology needed for SSP was developed in the late 1970s, at a time when an expected, inexpensive access to space via the space shuttle motivated concepts of large platforms in orbit. The early concepts for SSP were some of these. In the 1980s, NASA investment in construction of large space structures was largely funded and motivated by the problem of constructing early concepts of the International Space Station. This research was embodied in the 1987 Space Station Freedom baseline configuration.

One of the engineering points of view that developed at that time was that the most efficient structures for large space platforms would necessarily be assembled in orbit. It was recognized that, from the point of view of cost, mass, and packaged density, the best arrangement of material to form structures in space would probably be ones assembled and integrated there, rather than ones integrated on the ground and deployed. For example, the 1984 Reference Design for Space Station was completely deployable, but by 1987 NASA had changed the concept to a more mass-efficient assembled structure.

However, this point of view did not persist. When the assembly of the Space Station structure was abandoned in the early 1990s, NASA investment in this area largely stopped. The current "preintegrated" Space Station structure is, in fact, a multisegment assembly of launch-rated pieces. It is not a structure meant to function only in space. In this respect, unlike Freedom, the current ISS structure is not the kind of lightweight (some would say "gossamer") structure that SSP might require. Thus, although ISS is being assembled on orbit, the state of the art is arguably no closer to space construction now than 10 years ago.

In the intervening years, much of the focus within NASA has been on the development of highly efficient deployable structures. This has included inflated and lightweight unfurled structures, with a concomitant set of new concepts and materials. The SSP structures roadmap reflects this change in focus over the past decade. It would be useful, however, to reconsider the SSP structures roadmap in light of the basic goals and requirements of space construction. This may reinforce some of the decisions already made within the SERT program, but it may also lead to a selection of technologies different from those currently included in the SSP structures roadmap. The following section expands on this idea.

SSP Structural Design Requirements and Technical Issues

One basis for rational examination of the design of SSP structures is provided by Hedgepeth (1981) and Hedgepeth et al. (1978). The purpose of a structural and control subsystem is to provide a stable geometric shape, arrangement, and orientation of the components of the SSP. These requirements must be maintained under on-orbit loads. For SSP, like all large space structures, the loads include gravity gradient, pointing and slew accelerations, and thermal gradients. In addition, SSP loads include assembly, verification, and maintenance loads, which are loads not ordinarily considered design drivers in modern satellite structures.

The primary design objective for the structural subsystem is to meet the geometric requirements under the prescribed loads while minimizing cost. A space structure has several design influences, which for SSP can be separated as follows: (1) mass, (2) package density, (3) fabrication complexity, (4) integration and test, (5) lifetime, and (6) impact on other subsystems.

Mass

For SSP, without a doubt the total mass of the structure is an important component of the overall cost. The amount of mass for the concepts presented to this committee varied but was easily on the order of 1,000 equivalent shuttle payloads per SSP satellite (Olds, 2000; Carrington et al., 2000). Of this mass, the fraction allocated to the structure also varied but was on the order of 25 percent of the total mass of the SSP satellite.

The mass of the structural subsystem should not be estimated without due consideration of the requirements and specified loads. Different structures in space will have different masses, even if their size and shape are similar. As an extreme example, if a structure has no shape requirements, or if it experiences no loads, then it can have zero mass. The most "gossamer" of structures does nothing as a structure. So the structure must have finite mass in accordance with the requirements set forth in its design.

This actuality makes it difficult to extrapolate mass estimates for SSP from other space structures unless those structures satisfy the SSP requirements and loads. One should not, for example, extrapolate the mass of an SSP boom from the mass of a solar sail boom. According to material presented to this committee (Carrington, 2001; Canfield, 2001; Collins, 2000), some mass extrapolation was done to support two of the system analyses (the Abacus and the Integrated Symmetrical Concentrator concepts). For instance, "mass estimates for the inflatable rigidizable truss elements were derived from equations supplied by [a manufacturer]" (Carrington, 2001) and "mass elements for truss elements . . . were derived from [a Space Station prototype truss built at NASA Langley]" (Carrington, 2001). Some of the mass estimates were also based on the studies from the 1970s.

It is not asserted here that these mass estimates are wrong, but since they are based on extrapolations from the past, they may be misleading the specification of technological goals for the SERT effort. Without a top-down flow of the structural requirements, it is difficult to assess whether the stated technological goals are sufficient for SSP.

For large space structures such as SSP, it is recognized that the most important design criterion that drives the mass will be the stiffness required from the structure to withstand a given load and achieve a given required stability. When the loads are inertial, such as gravity gradient or attitude control accelerations, they vary with the structural mass. For this reason, it is the ratio of stiffness to mass that will drive the structural design of an SSP system.

This means that vibration frequency (the ratio of stiffness to mass), and not overall stiffness, is a primary structural design criterion for space structures. Usually the inertial loads will determine the frequency requirement. Sometimes, however, the pointing stability and bandwidth requirements for the attitude control system will determine it. Nevertheless, the first vibration frequency of the structure is often a top-level requirement levied on the structure (Hedgepeth, 1981).

This means that two structures for a given functional requirement or application can be compared with respect to how much mass they use to meet a first vibration mode requirement. In fact, one can look at the scaling of the first vibration mode as a basis for extrapolating structural mass (Lake, 2001; Lake et al., 2001). It is suggested here that the goals itemized in the SSP structures roadmap would be better supported if cast within such an overall design requirement.

Package Density

While mass is often considered the primary cost driver for the structure, the packaging density can have an important effect as well. If the mass is low but the structure still does not fit within the volume allocated by the booster, more launches will be required for the same amount of mass. It is for this reason that structural development goals should include the packaging density (kg/m^3) as well.

For example, it has been pointed out that existing space structures, whether communication antennas or the Hubble Space Telescope, are designed for packaged densities on the order of 65 kg/m^3 (Lake, 2001). This is close to the typical payload density of existing large boosters. This reference also points out that the trend in inflatable or unfurled concepts may not be meeting this requirement, meaning that those structures may in fact be volume critical, not mass critical, unless this requirement is factored into their development.

Fabrication Complexity

An SSP structure with a low mass that meets the packaging density requirement may still be too expensive to fabricate. That is to say, the number and complexity of the parts used in the structure can be an important design consideration. The number of parts might be a method for capturing this cost, measured in numbers of drawings for design time and fabrication time.

What this means, for example, is the best overall solution might be higher mass. In other words, it may be possible for a higher-mass structure to more than offset the cost of the extra launch mass if it is sufficiently simpler to design and build. Whether this is an issue for SSP should be examined (it is for current spacecraft). It is possible, however, that the mass production of SSP structures may offset this cost contribution.

Integration and Test

Perhaps the most challenging requirement for an SSP system is that it may be the first application to integrate and test the structure in orbit, perhaps the first application of space construction. It is standard practice today for space structures to be integrated and tested on the ground, then deployed on orbit. In fact, this activity is an essential risk mitigation activity. One must ensure that the structure will deploy reliably on orbit upon command. The need to mitigate this risk can add complexity to the system in the form of redundancy and in the form of repeated deployment and testing in 1 g.

An SSP structure, however, cannot be tested on the ground. The size of the structure would be difficult to accommodate. Moreover, when space structures must be integrated and tested on the ground, their design can be driven by the loads in 1 g, not the loads in 0 g. The inertial loads at GEO can be three to six orders of magnitude smaller than in 1 g, which means the first vibration mode required for an SSP structure will be smaller by the square root of the same factor. Integrating and testing an SSP system on orbit is a fundamental necessity.

However, integrating and testing structures in space (a.k.a. space construction) would also represent a fundamental change in engineering practice. Fundamental changes like this require not only new procedures, but also confidence born of success. This in turn requires that the procedures be proven in developmental applications and flight experiments. One major change that might be needed for SSP is the reliance on predictive models of structural performance once in orbit. Namely, ground-based testing may need to be replaced by analytical verification through simulations. Currently, in spite of advancements in computational capability, it is still not possible to predict the first vibration modal frequency or damping from fundamental mechanics with arbitrary accuracy. Damping, in particular, is almost always determined from tests, not from models. Either this needs to be remedied or the uncertainty in structural behavior needs to be accommodated in the design.

Lifetime

The lifetime of the structure is determined by both the degradation of material properties and the damage to the structure from orbital debris and maintenance activities. There are two consequences to this: (1) SSP systems will require long-lifetime materials and (2) SSP systems will require some level of structural repair and maintenance. The second concern may need to be considered in the development of the structural concepts. The repair or replacement of a truss member, for example, might not be possible if the structure cannot maintain overall integrity during the repair.

Impact on Other Subsystems

The structure often is at the center of the system design because other systems rely on its function to meet, in part, their own requirements. The development of structural technology should be factored into other SSP systems.

For example, the structural concepts should be developed in coordination with technology development in the robotic system. In fact, assembly and maintenance must be realized without inducing loads that become design drivers, or the mass of the structure may need to be increased. A mass-efficient SSP structure should not be designed for assembly loads.

Another possible synergy might be the use of PMAD components for part of the load-bearing structure (the notion of a multifunctional structure). The PMAD system represents in some concepts as much as 30 percent of the SSP mass, so use of a portion of this as structure might have a substantial impact on lowering the overall mass.

Challenges to Be Met

NASA provided no basis for establishing what goals are currently achievable with existing structural technology. It is not clear, for example, that an SSP system cannot be built using existing materials and controls technologies. Currently, goals in the structural area may be leading the NASA engineers to overdevelop structure for an SSP system. The following challenges are of concern: (1) requirements flow-down to structural stiffness and frequency requirements, (2) development of analysis-based verification techniques, and (3) development of modular structural concepts with low fabrication and assembly costs.

The need for requirements flowdown is perhaps very important to do early in the SSP program. A particular consequence of a structural requirements flowdown might be a refocus of planned R&D efforts. For instance, the required stiffness and mass of the concepts currently planned may be adequate for what is required; then again, they may be overly stiff. It is not the intent here to identify all possible impacts of such a reconsideration, but the following example illustrates the need for such an analysis.

The SERT roadmap calls for an overall system pointing accuracy of 3°. This can be used to develop basic requirements on the structure and the control system. It can be shown that a pointing accuracy of 3° for the solar array requires a controller bandwidth of 63 µHz (Hedgepeth et al., 1978). The concepts NASA presented to the committee have a structural natural frequency easily on the order of 100 times this bandwidth (0.007 Hz).

If the first mode of the structure is this far above the bandwidth of the attitude control system, it is unlikely that the attitude control will excite motion in the structure. Furthermore, it is also unlikely that structural vibrations will impact the gain margin of the controller, unless the damping ratio of the structure is smaller than what is typical in current space structures by a factor of more than 100. Thus, in this one example, it might be apparent that there is no need for a future SSP program to invest in control-structure interaction technology. Those funds might be more properly redirected to other areas. On the other hand, it might be that the structure is considerably stiffer than necessary and the mass required for the structure might be significantly reduced if this requirement were properly allocated to the structure.

Recommendations

The current structures and controls investment strategy is apparently not based on a quantitative analysis of the requirements for SSP. This produces two significant uncertainties for the SSP program. First, the structural mass estimates for the selected concepts may be considerably in error. Second, the real needs for the SSP structure and control may not be addressed by the selected technologies. Without this analysis, it is equally possible that either the selected technologies will fall short of the actual SSP system needs or SSP system needs might be met with no further investment in structural or control technologies. Further studies should be made to determine the structural requirements (strength versus flexibility, structural dynamic loads, mass, etc.) before large commitments of research funding are given to advanced *structural* technology development for the SSP program.

Also, based on the data presented to the committee, the SSP attitude control requirements should be met by a very-low-bandwidth attitude control system that avoids adverse interaction with the structural vibrations. Therefore it does not appear that there are any remaining technical challenges in structural *control* technologies and NASA might better use such funds elsewhere.

Particular attention should be given to (1) the derivation of structural mass scaling laws derived from SSP load and stability requirements; (2) an examination of the degree of interaction, if any, between the attitude controller and the structural vibrations; and (3) possible impacts of structural concept complexity on the fabrication cost.

Recommendation 3-7-1: The SSP program should perform a requirements-driven analysis of the overall structural mass, stiffness, and vibration frequency requirements needed for selected architectures. Technology goals, roadmaps, and budgets should be adjusted to support the results of this analysis.

Synergy with Other Programs

While it may be tempting to recognize synergy between an SSP system and civilian communication or military satellite structures, their structural requirements are radically different. In fact, one should be cautious about presuming without analysis that there is overlap in the technology needed by SSP and that being developed by other programs.

As discussed above, the frequency and stiffness required for an SSP system are very low, even compared with the requirements of current microwave geosynchronous satellites. While an SSP system might require only 3° of pointing control, Earth-pointing satellites might require much less than an arcsecond of error. The structural frequency requirements differ in proportion to the allowable error. This means SSP can have a vibration frequency perhaps 1/10,000th that of an Earth-pointing satellite.

3-8 THERMAL MATERIALS AND MANAGEMENT

Thermal Management Concepts

Thermal management is a critical spacecraft capability needed to maintain all spacecraft components and parts within a specified operational temperature range. Not only are spacecraft made of many dissimilar materials that expand and contract at different rates with respect to temperature, causing mechanical fatigue and eventual failure, but all electrical components have a specified operating temperature range needed to function properly. Without thermal management, a space-based system would quickly cease to operate. However, thermal management is generally one of the last technologies considered in spacecraft design. Thermal management encompasses everything from keeping components warm with heaters to dissipating excess heat via radiators. The dissipation of waste heat generally requires the greatest resources (mass, volume, and power) and includes collecting heat from electronic components (power converters, batteries, computer, rf antenna, etc.), transporting it away from these components, and rejecting it to the outside environment via a thermal radiator.

Current State of the Art

While major improvements have been made over the last decade in other areas of spacecraft bus technology (generation, storage, structures, etc.), minimal progress has been achieved in developing next-generation thermal management technologies such as heat pipes and deployable radiators. As a result, thermal management is rapidly becoming a limiting factor in designing larger, more powerful spacecraft, resulting in serious mass and volume penalties.

The state-of-the-art thermal management system is arguably the Boeing 702 spacecraft's deployable-loop heatpipe radiator, estimated to have a specific mass of about 4 kg/kW (Gerhart, 2001). Interestingly, the long-term (15-20 years) specific mass goal identified by the SERT program is 4 kg/kW. This figure greatly underestimates what is possible. A 4-kg/kW system would be far too heavy for systems larger than 25 kW. A new ultralightweight, flexible, deployable radiator has been developed by Creare, Inc., having a projected specific mass of ~1 kg/kW and areal mass density of 0.75 kg/m^2 using mechanical pumps and inflatable deployment structures presently under development.[5] Considering the present minimal investment in next-generation thermal management technologies, the metrics of 1 kg/kW and 0.75 kg/m^2 can be expected to achieve and remain the state of the art over the next 5-10 years.

Assuming that the heat generated by the SSP PMAD system is actively dissipated by the thermal management system—that is, 100 percent of the waste heat passes through an active cooling loop—then the resulting mass and size for the thermal management system would be formidable. For example, assuming that 3 GW of electrical power must pass from the solar array through the SPMAD electronics to enable 1.2 GW to reach the terrestrial utility power grid (the solar array must be oversized to account for SPMAD and WPT transmission and collection losses) and assuming the PMAD system is 90 percent efficient, then the heat generated by the SPMAD system would be 300 MW. The required thermal management system mass and area to support PMAD alone would be 3×10^5 kg and 4×10^5 m^2, respectively, requiring approximately 12 flights of the current space shuttle at maximum payload capacity to simply lift the PMAD components to LEO. However, it is noted that a significant reduction in thermal control mass could be achieved if the SSP PMAD system was distributed throughout the large area of the solar array, potentially enabling passive cooling via heat radiation directly out to space.

Technical Performance Goals Needed for Economic Competitiveness

Successful development of an economically viable SSP system will require large advances in development of space solar array, PMAD, thermal control, wireless transmission, and launch technologies. Improvements in thermal management technologies alone will not enable SSP to be economically competitive with terrestrial utility electricity. However, significant thermal management mass and cost savings could be achieved via use of heat pumps to increase radiator temperature. The size and mass of the radiator is determined by the maximum temperature of the radiator; the greater the temperature, the smaller and lighter the radiator. The 1-kg/kW and 0.75-kg/m^2 performance metrics above are based on a rejection temperature of 120°C driven by today's maximum PMAD operating temperature. It is projected that heat pump technology could be available in the 2010-2015 time frame, allowing an increase in radiator temperature to as high as 300°C, enabling a reduction in power density to 0.4 kg/kW (Gerhart, 2001). The SERT program did not address the potential impact of future heat pump technology on thermal management performance.

Challenges to Be Met

As stated above, improvements in thermal management technology alone will not enable SSP to be economically competitive with terrestrial utility electricity. The committee believes that the greatest challenge to support SSP is to develop more realistic and accurate SSP system cost and performance models that include theoretical thermal management, solar array, PMAD, and wireless transmission cost and performance parameters, to be able to realistically quantify SSP mass, volume, cost, and launch challenges.

Recommended Priority for Investment

There is no single or simple solution for reducing the large size and mass of the thermal management system required to dissipate SSP PMAD system waste heat. In general, what is required is development of higher operating temperature PMAD electronics and thermal management components. To support higher operating temperatures, advanced heat pump development should be accelerated.

Recommendation 3-8-1: The SSP program should focus efforts in thermal management and materials on revolutionary high-payoff, high-risk approaches having potential cost and mass savings not only for

---

[5]Research was funded on this radiator through a U.S. Air Force Small Business Innovation Research program entitled "Freeze-Tolerant, Lightweight, Flexible Radiator," Contract Number F29601-98-C-0115.

SSP-related activities but also for most other space technology development efforts.

Recommendation 3-8-2: The goals of the SSP program in the thermal management and materials area should be re-evaluated, taking into consideration the current state of the art in thermal management and projections of technology advancement in the next 10-20 years.

Synergy with Other Programs

Thermal management is considered a pervasive spacecraft technology and important for all near- and far-term DOD, NASA, and commercial space programs. Thermal management is also critically enabling for all DOD, NASA, and commercial aircraft. Any performance improvements in thermal management technology, such as reduced specific mass (kg/kW) and areal mass density (kg/m^2), will greatly benefit a multitude of future government and commercial space and aircraft systems.

3-9 SPACE TRANSPORTATION AND INFRASTRUCTURE

Space transportation will be a vitally important aspect of a commercial-scale SSP initiative. Past studies have shown that a major portion of the total deployment cost for this program will be attributable to this one area (Davis, 1978). Two space transportation activities are now expected: (1) launch of the massive elements of an SSP system from Earth into LEO and (2) subsequent transfer of parts or partly assembled elements of an SSP system into the operational GEO location currently favored by the program because of its inherent orbital properties, lower levels of orbital debris, and room for assembly and maintenance activities. For an operational program, large reusable space launch vehicles are expected to serve the former needs, while highly efficient electric propulsion may be used for orbit transfer.

Launch Vehicle Needs

NASA demonstrated that in order to reduce costs for the system as a whole, vehicles and technology useful for continuous build and launch capability would be necessary. The scale of space launch activities necessary to place meaningful capacity for supplying Earth's needs for electrical power was illustrated to the committee as follows (Olds, 2000): 450 flights per year for 30 years of a launch vehicle capable of delivering 40 tons per flight or over 18,000 tons (40 million pounds) per year. This level of activity is 80 times that of current space launch activity. For example, eight flights per year of today's space shuttle will produce useful mass in LEO of about 160 tons. If other space launch vehicles deliver half this amount, the multiplier of 80 is confirmed.

If future studies and demonstrations indicate qualitative and cost viability of space power, two important questions will be the rate and the timing of penetration of the global market for this new source of electrical energy. If SSP is shown to be sufficiently economical to begin such a program, the demand to place these satellites may rapidly grow well beyond that indicated in the NASA SERT studies. Other large-scale uses of space by both the global private and public sectors may occur. These new uses are expected by many analysts to be contemporary with an SSP system and will multiply the levels of launch activity associated with it (Andrews and Andrews, 2001).

No one can accurately forecast today if these huge increases in traffic to space will be realized. Similarly, although extrapolations and estimates may be made, no one can accurately determine, today, the costs of space transportation 30 and more years in the future. As a result, decisions on the viability of future space solar power efforts should not be based solely on the space transportation segment of the program.

If today's expendable launch vehicles persist for another four decades and if present forecasts made by their suppliers' of an inelastic, or fixed-size, launch market prove to be accurate, continued high launch costs will render power from space uneconomical. If, however, new concepts and approaches for space launch are adopted and evolved and if new space launch traffic is attracted by lower costs and higher reliability, then costs will fall, perhaps quite dramatically.

Future space transportation work within any near-term SSP program should be focused on the needs of a major NASA SSP demonstration project (i.e., MSC 3). Such a program will require significant space transportation. If a 10-MW full-scale demonstration is selected and if the specific mass is that of earlier NASA studies, the satellite may require 150 to 200 tons of mass placed into orbit. If the demonstrator satellite is to be placed into GEO, additional mass in LEO will be necessary. Extended experimental use of this demonstration unit is likely, further increasing the space transportation needs of this program. Thus, the MSC 3 SSP demonstration satellite program may produce a space launch task approaching the magnitude of placing the International Space Station into orbit.

During the same interval, NASA and others may embark on ventures in space as large or larger than this demonstration of power from space. These may include establishing permanent bases on the Moon, human exploration missions to the Martian moons or the surface of Mars, exploitation of asteroid resources, or public space travel.

The NASA SERT program has established, for a commercially viable SSP system, a goal of $400/kg of mass as the allowable price for each of the two transportation legs—Earth to orbit (ETO) and in space (orbit to orbit). Determining whether this can be achieved is not within the scope of the present assessment; however, the committee feels that this goal may not be low enough to support a commercially viable SSP terrestrial power system. The costs of space transportation must be reduced below $800/kg if the present cost goals of the NASA SERT program are to be realized. Mass goals must also be tightened for other SSP subsystems since these will determine the size of the transportation task. Re-evaluation of these goals is important to SSP strategy because it will influence the technologies that should be pursued. Current space transportation goals are already driving near-term choices within the SERT program, creating the possibility that technology investment choices may be made based on goals that are not stringent enough in this area.

Recommendation 3-9-1: The SSP program should re-evaluate its cost goals in the space transportation area. NASA should ensure that its space transportation technology development activities consider the revised SSP goals.

In-Space Transportation Needs

One important driver in establishing on-orbit SSP cost is in-space transportation (i.e., LEO-to-GEO transportation). The current SERT program has not yet prioritized investments in this area, although budget schedules for the program show increased activity in this area by 2006. Low-cost options for in-space transportation may require large advancements in technology—advances that may not occur within the next 5-10 years without program stimulus. As a result, investments need to be made in this area, whether under the purview of the SSP program or as a separate NASA initiative. Applications in addition to SSP will be available for effective use of such new in-space transportation options.

Technologies for in-space transportation such as electric, solar-electric, magnetohydrodynamic, ion, and solar-thermal propulsion could all be considered as possible candidates. Low cost and high rate of usage will be driving factors for SSP since multiple use of vehicles will be required. Preliminary cost analyses by the SERT program have also shown that costs related to ownership and use of in-space transportation, as well as launch vehicles, will have impact on the commercial viability of terrestrial power from space. Detailed trade studies of various in-space transportation concepts should be made along with satellite design and assembly concepts in order to establish lowest cost methods for placing SSP elements in GEO. A preliminary assessment of these needs and ideas to meet those needs was performed during the SERT program through a university-funded research grant. Again, as for launch vehicles, cost and mass goals must be tightened because they will influence the technologies that should be pursued.

Recommendation 3-9-2: Detailed trade studies for various in-space propulsion options should be performed early in the program. Issues involving cost, packaging, mass, trip time, and risk assessment must be incorporated into the vehicle choice.

Flight Test Demonstration Transportation Needs

Although not dependent on major advances in space transportation within the next 5 years, planned SSP flight test demonstrations will require detailed assessment of needs in the area of space transportation. The committee understands that configurations have not yet been chosen for the future SSP technology flight demonstrations (TFDs). However, planning must begin now to determine if current launch vehicles will be adequate and if sufficient mass and volume are available on the shuttle or suitable expendable launch vehicles in the time frame anticipated.

Other than very preliminary mass estimates for various TFD configurations, little systems work has been seen by the committee in the area of space transportation. Most space transportation analysis within the program has been concentrated on space transportation needs for the full-scale MSC 4 system. This is an important factor in determining the appeal of a commercial terrestrial SSP system; however, launch vehicle and in-space transportation needs for the early MSCs must be assessed during the planning stages.

Recommendation 3-9-3: The SSP program should develop detailed space transportation requirements for all technology flight demonstrations (MSC 1, MSC 1.5, and MSC 3) that include data from studies on packaging, cost and mass estimates, and other important parameters.

Opportunities for Synergy

NASA's efforts toward the new Space Launch Initiative are intended to reduce risk for NASA's future needs, with funding of $290 million to NASA in FY 2001 and plans by NASA to expend up to $5 billion in this area during the next 5 years. However, nowhere in the descriptions of the SLI provided to the committee is power from space mentioned (Davis, 2000).

Improved coordination is needed between NASA space transportation efforts and the future needs of the United States for space launch, including any future SSP program. Work should be conducted to explore the ramifications of other new markets and to describe the evolutionary paths that might bring this low-cost launch capability into operation. Information should be provided to SLI on the SSP program's needs for future TFDs, including cost goals, optimal payload mass, packaging, launch rates, reliability needs, and scheduled need dates to achieve the SSP goals.

Recommendation 3-9-4: The SSP program should begin discussions between its management and that of the NASA Space Launch Initiative, so that future milestones and roadmaps for both programs can reinforce one another effectively. This discussion should include specific information on SSP space transportation needs, including cost goals, timelines for deployment, optimal payload mass, packaging requirements, launch rates, and reliability requirements.

NASA's Advanced Space Transportation (AST) program (NASA, 1999) was created to achieve NASA's goal of significantly lower space transportation costs. The program has initiatives in four major areas, including the support of long-term technology research in advanced chemical and nonchemical propulsion systems for use in space. Advancements in the areas of high-power electric propulsion (Hall and ion thrusters), cryogenic engines, spacecraft miniaturization, solar-thermal powered transfer, electrodynamic tether orbit transfer, and in-situ propellant utilization are expected under the program. Propulsion research is also being supported in the advanced concept areas of magnetic levitation, pulse detonation, beamed power, magnetohydrodynamics (MHD), fusion, antimatter annihilation, and breakthrough physics. Currently, there is little evidence of interaction between the SSP program and the AST program in developing new in-space transportation options.

Other NASA efforts not labeled specifically within the AST program are also being made in electric and solar electric propulsion, including programs at several NASA centers and the Jet Propulsion Laboratory. Many of these efforts are in conjunction with academic institutions. Efforts in NASA also include advanced chemical and MHD propulsion, beamed energy and momentum propulsion, and fission and fusion propulsion. NASA co-sponsors, in conjunction with the University of Alabama in Huntsville, an annual joint conference on advanced space propulsion concepts[6] from which current technology advances can be identified and the efforts of the academic community better utilized.

Recommendation 3-9-5: The SSP program should encourage expansion of the current in-space transportation program within NASA and interact with its technical planning to ensure that SSP needs and desired schedules are considered.

Currently, many other non-NASA investments are being made in orbit-to-orbit transportation research, including major programs within the Air Force and industry. There is little evidence of the SERT program leveraging these program investments or partnering with internal NASA space transfer technology programs. The Air Force Research Laboratory's Propulsion Directorate currently has research programs in electric propulsion, solar thermal propulsion, pulsed

---
[6]Proceedings for the last three meetings are available online at <http://www.eb.uah.edu/maglev/aspw/>. Accessed on August 16, 2001.

plasma propulsion, and Hall effect thrusters.[7] The U.S. Navy Research Laboratory's Naval Center for Space Technology also supports research efforts in the areas of MHD, tether, and electric propulsion technology.[8]

Recommendation 3-9-6: The SSP program should increase coordination of industry, academic, and other NASA and non-NASA government investments in advanced in-space transportation concepts, particularly in the areas of electric, solar-electric, magnetohydrodynamic, ion, and solar-thermal propulsion.

3-10 ENVIRONMENTAL, HEALTH, AND SAFETY FACTORS

Objective and Scope

This section considers the environmental, health, and safety issues of SSP systems. Environmental factors include any significant effects on the space surrounding the satellite in orbit, on the media through which power is transmitted, and on the neighborhood of the receiving station. These effects generally include contamination of the environment in the form of communications interference, generation of debris in orbit, or possible effects on the atmosphere (i.e., power transmission to Earth receiving sites). Safety factors include any health hazards to Earth biota associated with power beaming, whether in the form of microwave or laser transmission. All of the concerns associated with environmental, health, and safety factors must be addressed early in the program, with particular emphasis on public awareness and public perception. If any of these risks are found to be of significant public concern, the public may be reluctant to host an SSP ground receiving station regardless of the economic competitiveness of SSP. Environmental, health, and safety risks may be minimized through appropriate technology investments and technology design guidelines. However, higher priority should be placed on these areas by the SSP program.

In designing a full-scale SSP system, the environmental impact analysis must also include pollution that may occur during emplacement of the facility (production and launch of transport vehicles), as well as environmental effects during the operational phase. However, it is also essential to include, in the case of SSP for terrestrial power generation, positive effects such as decreased pollution in comparison with other forms of power generation (i.e., fossil fuels). The great appeal of terrestrial SSP, of course, is that, once in place and operating, its contribution to greenhouse gases in the atmosphere is zero.

Current Issues

Environmental, health, and safety issues are now recognized as essential concerns to be addressed as early in a program as possible. Some 60 percent of the effort in the early solar power satellite studies was devoted to investigating environmental and societal issues (Koomanoff, 2001). During the SERT program, a working group was assembled to look at environmental, health, and safety factors. Preliminary findings of this group show that the following are likely to be issues (Anderson, 2000):

- Lack of information on the risks associated with exposure to low-level microwaves, particularly long-term human exposure to low-level microwaves;
- Lack of data on effects of laser beaming options on humans;
- Spectrum allocation issues (international considerations);
- Orbital space allocation issues (in geosynchronous orbit);
- Land use availability for receiving systems for terrestrial SSP;
- Transmission interference concerns (both on orbit and on the ground); and
- Possible use of SSP systems as a weapon.

The effects of power beaming are primarily a function of the frequency and the energy density of the beam. Currently, the frequencies and energy levels for SSP power transmission are estimated to be as follows: microwave, 5.8 GHz (5.2-cm wavelength) and 1 kW/m^2 power level; and laser, 1.03-mm wavelength and a yet-to-be-determined kW/m^2 (note: design will be eye safe due to distributed transmitting sources).

These values represent very low energy levels, and

---

[7]More information on the Air Force Research Laboratory's space propulsion programs is available online at <http://www.pr.afrl.af.mil/>. Accessed on August 16, 2001.

[8]More information on the Naval Center for Space Technology's programs in space propulsion is available online at <http://ncst-www.nrl.navy.mil/>. Accessed on August 16, 2001.

in both cases, the beam will be designed to be quickly "turned off." Microwave transmissions can be dephased, dramatically reducing energy density. Laser strengths will be planned only at eye-safe levels. Current design specifications for any future SSP system allow only low energy densities and design-safe standards to prevent further focusing of the beam. The center of the beam's Gaussian distribution is planned for energy densities of about 23 mW/cm^2, or about one-fifth the intensity of summer sunlight at noon (Mankins, 2000a; Moore, 2000). Additionally, any residual energy outside the rectenna's protective fence would be far below current microwave safety standards, between 0.01 and 1 mW/cm^2 (Moore, 2000).

Most research in the area is focused on the effects on humans, animals, and biota of radiation from household devices, digital phones and other electronic equipment, and electric utilities (NIEHS, 1999). Little research has been performed at field levels specific to SSP application. However, research has been done on bees and birds exposed to microwave radiation at twice the dose expected for a creature flying through a typical microwave power transmission beam. Results to date indicate that there is no effect, at least on the animal's directional flying ability (Koomanoff, 2001). Other testing has been performed on monkeys and is now under way with humans exposed to low-level microwave radiation. Results to date from this testing indicate that such exposure apparently does not render the subject sterile or result in cataracts or any other deleterious effects (Kolata, 2001). The larger problem may be preventing animals from roosting on the rectenna or damaging portions of the receiving site—issues that are already of concern to terrestrial solar power generation farms.

Research was previously funded on the effects of microwave radiation on plants as well (Michaelson and Lin, 1987); however, only a small body of work has been performed or published in this specific area (Skiles, 2001). Results from the previous research have been inconclusive. Studies have shown that microwaves (i.e., 2.45 GHz) generally have an inhibiting effect on plant growth (Picazo et al., 1999) and that reduced growth rates were demonstrated for higher power densities (Urech et al., 1996). Other studies show that effects of radio frequency waves may vary with the stage of plant development (Magone, 1996) and that no significant differences between exposed and unexposed cells were seen for plants exposed to 41.6 GHz frequency (Gos et al., 1997). The SERT program is currently funding at least one experiment to determine the effects of microwave fields on alfalfa (Skiles, 2001). The committee believes that further studies in the areas of environmental, health, and safety are warranted.

One of the topics currently under investigation is determining how much small transmitters such as cellular phones may interfere with aircraft avionics and communications. Results of these studies may indicate whether there is reason for concern about aircraft flying through an SSP power beam. Many different industries and government institutions are also performing research to determine if microwave transmitters can cause cancer. So far, results appear to show that no correlation exists, but public perception is driving cell phone manufacturers to change their designs even without clear scientific evidence. The SSP program may well be subject to the same sort of public relations requirements. The committee understands that the relevance of such cellular phone research to space solar power may be questionable. However, program managers must be aware that the long-term effects of SSP wireless power transmission on humans *must* be quantified before public acceptance is found.

As mentioned previously, a major issue is the lack of research on the environmental, health, and safety effects of exposure to this WPT system. Technologists must be able to prove very limited risk during use. Interference effects of the chosen WPT method on satellites and other spacecraft in the vicinity of the SSP system should be investigated as well.

Concerns are very real about the effect of the orbital debris environment on large space structures in LEO (NRC, 1995; NASA, 2000). The problem is not just the present orbital debris population, but the worsening evolution of this debris population, increased perhaps by SSP operations. The orbital debris pollution in both LEO and GEO will remain and worsen. Research is currently under way to model the orbital debris environment more accurately and predict its increase in time. In addition, testing of hypervelocity impacts on spacecraft materials is ongoing. Also, methods of collision avoidance are being developed to aid spacecraft in avoiding collision with mid- to large-sized debris that can be tracked.

Technical Goals Needed for SSP

Systems studies should be done to clearly show not only that the SSP system is safe but also that the overall environmental impact of a space solar power system is a positive one. Although once in operation the SSP system essentially generates power with zero pollution, the pollution produced during the construction and deployment of an SSP system (i.e., multiple space launches) must be within acceptable levels. This study of balance would have to include, for example, any environmental pollution that occurs during emplacement of the facility, such as harmful exhaust from the large number of launches necessary to transport the components of the system to orbit (currently the SERT program estimates between 500 to 900 Earth-to-LEO launches, each with a 40-metric-ton capacity, for one 1.2-GW SSP facility, depending on the chosen design).

Environmental effects of the SSP power beam on the atmosphere, on Earth biota, and on the space environment surrounding the SSP satellite will clearly have to be minimal in all respects. It is not sufficient to simply show scientifically that effects are likely to be minimal; it is essential to mitigate from the outset any and all perceptions that the system may be harmful or dangerous in any way to humans, animals, or the environment. Politics and public opinion will weigh in heavily on the ultimate viability of SSP systems. Issues of electromagnetic compatibility between various hardware components must also be treated during the course of technology development. As technology readiness is improved, studies should be performed on individual technologies to determine the level of electromagnetic compatibility between the SSP system components themselves and between the wireless power transmission system and external electronic hardware such as avionics, terrestrial wireless communications systems, communications satellites, and medical devices.

It will also be essential to show that the system cannot be used as a weapon. Fail-safe features should be designed into the system to prevent pointing a full-strength beam anywhere but at the receiving site to which it is designated. Current plans are to keep beam density low. For the laser option, for example, a person standing in the center of the receiving site should be able to look up at the beam and experience no eye or skin damage. In addition, to assuage any fears that this technology could be reworked to be used as a weapon, the program should be international in structure, with full disclosure of information to all participating countries so that no strategic advantage can be gained by any one nation.

Challenges to Be Met

In terms of environmental, health, and safety factors, the greatest challenges for SSP are to develop a system with benefits that outweigh the costs and a high level of safety that can be proven to the public. These challenges can be addressed by funding near-term and continuing research on the topics listed in the following two sections at frequencies, power levels, and orbital locations specific to SSP. The lists are neither exhaustive nor prioritized.

Environmental Challenges

The following topics are areas of environmental research that must be investigated before deployment of a full-scale SSP system (by NASA, other appropriate organizations, or industry):

- Quantify the net environmental effect of having terrestrial SSP available in comparison with reliance on conventional sources of energy, including the polluting effects of building and launching SSP hardware and the possible deleterious effects of power beaming through the atmosphere.
- Develop construction methods for SSP systems that minimize the generation of additional orbital debris.
- Provide for efficient and final disposal methods for solar power platforms once their useful life is over; otherwise, they too will contribute to the orbital debris problem not just in LEO but in GEO as well (Anderson, 2000).
- Determine how SSP systems will alter the radiation environment around them and quantify the resulting impact in terms of orbital space allocation and interference effects at GEO.
- Assess power transmission side-lobe effects (energy transmitted outside the nominal beam diameter) and the impact on communications in the vicinity of the power beam as well as around the receiving station (ground based or in orbit).
- Perform additional research on dual use of rectenna sites for terrestrial solar power application, so that sufficient real estate can be found in all potential locations.

Health and Safety Challenges

The following topics are areas of health and safety research that must be investigated before deployment of a full-scale SSP system (by NASA, other appropriate organizations, or industry):

- Perform further research on the effects, especially long-term ones, of low-level microwave and laser radiation on Earth biota (humans, other animals, and plants).
- Design the SSP system so that use as a weapon is impossible under any circumstance.
- Establish public awareness and education outreach programs covering the benefits of SSP, the technology behind it, and the built-in safeguards. Such programs might include development of demonstration models (Pignolet, 2001), a children's book on SSP, or op-ed pieces on space solar power.

Recommendations

Recommendation 3-10-1: The SSP program should expand its environmental, health, and safety team in order to review SSP design standards (beam intensity, launch guidelines, and end-of-life policies); assess possible environmental, health, and safety hazards of the design; identify research if these hazards are not fully understood; and consider legal and global issues of SSP (spectrum allocation, orbital space, etc.). One approach would be to involve an international organization such as the International Astronautical Federation Space Power Committee in such studies.

Recommendation 3-10-2: Public awareness and education outreach should be initiated during the earliest phases of an SSP program to gain public acceptance and enthusiasm and to ensure ongoing support through program completion.

Recommendation 3-10-3: The SSP program should collaborate more effectively with other NASA programs in space-related environmental, health, and safety and with external industry and government agencies currently performing research related to such issues. Funding should be increased to cover gaps in research, specifically in areas that overlap SSP-related technology. NASA should initiate research that evaluates the effects of microwave and laser wireless power transmission at levels planned to be utilized by a full-scale SSP system.

Recommendation 3-10-4: Studies should be initiated to determine the effects of the electromagnetic incompatibility of wireless power transmission systems and avionics, terrestrial wireless communication systems, and SSP-related electronics.

Synergy with Other Programs

Research is currently being conducted at NASA and DOD on possible strategies to address orbital debris in both LEO and GEO (NSTC, 1995). These strategies include radar tracking and de-orbit stabilization, among others. Programs are also in place in fail-safe beam pointing technology and wireless power transmission. The Federal Communications Commission will be concerned about interference of radio frequency beams with current and planned communications systems and issues of spectrum allocation. Before deployment of a full-scale system, international issues in spectrum allocation, orbital space allocation, and orbital debris evolution must also be considered.

Research is being conducted by the wireless communications industry, the federal government, and other interested parties to settle questions about whether microwave transmission can be harmful to humans. The Electric Power Research Institute currently has a substantial multidisciplinary program that evaluates all aspects of health, risk, and field management research in respect to electricity and power.[9] This program includes efforts in epidemiology, toxicology, engineering, and other sciences. The National Institute of Environmental Health and Safety convened a working group on the effects of electromagnetic waves and currently has in place a program of public information and dissemination on the effects of electric and magnetic fields (NIEHS, 1999). Although frequencies and beam strength are different, inclusion of representatives from these organizations in any future environmental working groups would be beneficial. The SSP program should also capitalize on expert knowledge through

---

[9]More information on the Electric Power Research Institute's research in environmental, health, and safety issues of electric power is available online at <http://www.epri.com/>. Accessed on August 15, 2001.

work with the Committee on Space Research established by the International Council for Science—in particular, its work on environmental, health, and safety issues. The group should also continue its work with the International Astronautical Federation Space Power Committee.

3-11 PLATFORM SYSTEMS

Objectives and Scope

This subsystem of the SSP system consists of all the additional systems that are needed to construct a complete, integrated spacecraft, including attitude control sensors (Sun sensors, star trackers, and gyros), flight computers, the telemetry and data handling system, the command system, communications, and so on.

Current State of the Art

All of the above-mentioned subsystems currently exist on most spacecraft. The state of the art is advanced, and further advances will be made independent of the SSP application.

Technical Performance Goals Needed for Economic Competitiveness

The aforementioned subsystems make up a negligible fraction of the mass and cost of the SSP system. Therefore, economic competitiveness is not affected by these subsystems. Autonomous robots will place the highest demand on space computing.

Challenges to Be Met

There are many improvements possible in various satellite subsystems technologies. However, any advancements will be driven by many other applications.

Recommended Priority for Investment

Recommendation 3-11-1: The SSP program should not expend any resources to develop advanced technologies for standard satellite subsystems (e.g., attitude control sensors, flight computers, telemetry and data handling systems, and communications systems).

Synergy with Other Programs

There are many other programs in NASA and DOD to improve the performance of these subsystems with which the SSP program could collaborate.

REFERENCES

Akin, Dave, and Russ Howard. 2001. Personal communication from Dave Akin and Russ Howard, University of Maryland Space Systems Laboratory, to Mary Bowden, University of Maryland Space Systems Laboratory, February.

Akin, D.L., M.L. Minskshy, E.D. Thiel, and C.R. Kurtzman. 1983. *Space Applications of Automation, Robotics, and Machine Intelligence Systems (ARAMIS)—Phase II*, NASA CR-3734, 3735, and 3736, October 1983.

Anderson, John. 2000. "Environmental and Safety Factors." Briefing by John Anderson, NASA Marshall Space Flight Center, to the Committee for the Assessment of NASA's Space Solar Power Investment Strategy, National Academy of Sciences, Washington, D.C., September 13.

Andrews, Jason, and Dana Andrews. 2001. *Future Space Transportation Study*. Prepared for NASA Marshall Space Flight Center under contract No. NRA 8-27 TA1.1. El Segundo, Calif.: Andrews Space and Technology.

Ashley, Chris. 2001. Personal communication from Chris Ashley, Boeing Satellite Systems (formerly Hughes), to Rhoads Stephenson, retired, Jet Propulsion Laboratory, February 13 and April 5.

Bower, Ward. 2001. Personal communication from Ward Bower, Sandia National Laboratories, to Rhoads Stephenson, retired, Jet Propulsion Laboratory, January 31.

Canfield, Stephen. 2001. *Selected Results from Design of Structural Elements for the Integrated Symmetrical Concentrator*. Personal communication from Stephen Canfield, Tennessee Technological University, to Karen Harwell, National Research Council, January 3.

Carrington, Connie. 2001. *Response to NRC Structures Question (#12)*. Personal communication from Connie Carrington, NASA Marshall Space Flight Center, to Karen Harwell, National Research Council, January 3.

Carrington, Connie, and Harvey Feingold. 2000. "SERT Systems Integration, Analysis, and Modeling." Briefing by Connie Carrington, National Aeronautics and Space Administration, and Harvey Feingold, Science Applications International Corporation, to the Committee for the Assessment of NASA's Space Solar Power Investment Strategy, National Academy of Sciences, Washington, D.C., September 13.

Collins, Tim. 2000. *Solar Array Support Structure: Configurations and Analysis*. Presentation to the National Aeronautics and Space Administration, June 16.

Culbert, Chris, Brett Kennedy, William Whittaker, Peter Staritz, and Han Thomas. 2000. "Space Solar Power Robotics Assembly, Maintenance, and Servicing." Briefing by Chris Culbert et al., National Aeronautics and Space Administration, to the Committee for the Assessment of NASA's Space Solar Power Investment Strategy, National Academy of Sciences, Washington, D.C., September 14.

Culbert, Chris, Brett Kennedy, Hans Thomas, William Whittaker, and Peter Staritz. 2001. "Space Solar Power Robotics, Assembly, Maintenance, and Servicing." Briefing by Chris Culbert, Brett Kennedy, Hans Thomas, NASA; and Red Whittaker and Peter Staritz, Carnegie Mellon University, to the Space Solar Power International Forum, Washington, D.C., January.

Davis, Danny. 2000. "2nd Generation RLV Summary". Briefing to the Committee for the Assessment of NASA's Space Solar Power Investment Strategy, National Academy of Sciences, Washington, D.C., October 24.

Davis, Hubert. 1978. *Solar Power Satellite: Power from Space . . . a New Opportunity*. Paper presented to the 71st Annual Meeting of the American Institute of Chemical Engineers, Miami Beach, Florida, November 13.

Dickinson, Richard. 2000. Personal communication from Richard Dickinson, Jet Propulsion Laboratory, to Rhoads Stephenson, retired, Jet Propulsions Laboratory, December.

Erb, R.B. 2000. "Interest and Activities in Space Solar Power Outside the USA." Briefing by Bryan Erb, Canadian Space Agency, to the Committee for the Assessment of NASA's Space Solar Power Investment Strategy, National Academy of Sciences, Washington, D.C., December 14.

Feingold, Harvey. 2000. "SERT Systems Integration, Analysis, and Modeling." Briefing by Harvey Feingold, Science Applications International Corporation, to the Committee for the Assessment of NASA's Space Solar Power Investment Strategy, National Academy of Sciences, Washington, D.C., October 23.

Gerhart, Charlotte. 2001. Personal communication from Charlotte Gerhart, Air Force Research Laboratory, Space Power Group, to Kitt Reinhardt, Air Force Research Laboratory, January.

Gos, Pascal, Bernhard Eicher, Juerg Kohli, and Wolf-Dietrich Heyer. 1997. "Extremely High Frequency Electromagnetic Fields at Low Power Density Do Not Affect the Division of Exponential Phase Saccharomyces Cerevisiae Cells." *Bioelectromagnetics* 18(2):142-155.

Guha, Shubinda, Jeff Yang, Paul Nath, Jeff Call, and Tom Glatfelter. 1999. *Low Cost and Lightweight Amorphous Silicon Alloy Solar Array for Space Applications*. 34th IECEC Conference Proceedings, Paper No. 1999-01-2553, Vancouver, B.C., August.

Hazelrigg, George. 1977. Space-Based Solar Power Conversion and Delivery Systems Study. Report prepared for the National Aeronautics and Space Administration under Contract No. NAS8-31308, Princeton, N.J.: ECON, Inc.

Hazelrigg, George. 1992. "Cost Estimating for Technology Programs." In *Space Economics—Progress in Astronautics and Aeronautics*, Vol. 144, J.S. Greenberg and H.R. Hertzfeld, eds. Washington, D.C.: American Institute of Aeronautics and Astronautics.

Hazelrigg, George, and Joel Greenberg. 1991. "Cost Estimating for Technology Programs." Paper presented at the 42nd Congress of International Astronautical Federation, Montreal, Canada.

Hedgepeth, John M. 1981. *Critical Requirements for the Design of Large Space Structures*. NASA CR 3484. Washington, D.C.: National Aeronautics and Space Administration.

Hedgepeth, John M., Martin M. Mikulas, and Richard H. MacNeal. 1978. "Practical Design of Low-Cost Large Space Structures." *Aeronautics and Astronautics*, October, pp. 30-33.

Hoeber, C.F., E.A. Robertson, I. Katz, V.A. Davis, and D.B. Snyder. 1998. *Solar Array Augmented Electrostatic Discharge in GEO*. AIAA Paper 98-1401. Reston, Va.: American Institute of Aeronautics and Astronautics.

Kolata, G. 2001. "Tuning In to the Microwave Frequency." *New York Times*, January 16. Available online at <http://www.nytimes.com/learning/students/pop/010117wodwednesday.html>. Accessed on August 16.

Koomanoff. 2001. Personal communication from Fred Koomanoff, Sunsat Energy Council, to Mary Bowden, University of Maryland, Space Systems Laboratory, January 23.

Koproski, Ben, and Bob McConnell. 2001. Personal communication from Ben Koproski and Bob McConnell, National Renewable Energy Laboratory, to Rhoads Stephenson, retired, Jet Propulsion Laboratory, February.

Kurtz, Sarah, Dave Myers, and Jerry Olsen. 1997. "Projected Performance of 3- and 4-Junction Solar Cell Devices Based on GaInP and GaAs." Pp. 875-880 in *Proceedings of the 26th IEEE Photovoltaics Specialists Conference*.

Lake, Mark. 2001. "Launching a 25-Meter Space Telescope: Are Astronauts a Key to the Next Technically Logical Step After NGST?" Paper presented at the 2001 IEEE Aerospace Conference, Big Sky, Montana, March 10-17.

Lake, Mark, Lee D. Peterson, and Marie B. Levine. 2001. "A Rationale for Defining Structural Requirements for Large Space Telescopes." AIAA Paper 2001-1685 in *Proceedings of the 42nd Structures, Structural Dynamics and Materials Conference*, Seattle, Washington, April. Reston, Va.: American Institute for Aeronautics and Astronautics.

Magone, I. 1996. "The Effect of Electromagnetic Radiation from the Skrunda Radio Location Station of Spirodela Polyrhiza (L.) Schleiden Cultures." *Science of the Total Environment* 180(1):75-80.

Mankins, John. 2000. "Space Solar Power Exploratory Research and Technology (SERT) Program Status." Briefing by John Mankins, National Aeronautics and Space Administration, to the Committee on the Assessment of NASA's Space Solar Power Investment Strategy, National Academy of Sciences, Irvine, Calif., December 14.

Mankins, John, and Joe Howell. 2000a. "Space Solar Power (SSP) Exploratory Research and Technology (SERT) Program Overview." Briefing by John Mankins and Joe Howell, National Aeronautics and Space Administration, to the Committee for the Assessment of NASA's Space Solar Power Investment Strategy, National Academy of Sciences, Washington, D.C., September 13.

Mankins, John, and Joe Howell. 2000b. "Strategic Research and Technology Roadmap." Briefing by John Mankins and Joe Howell, National Aeronautics and Space Administration, to the Committee for the Assessment of NASA's Space Solar Power Investment Strategy, National Academy of Sciences, Irvine, Calif., December 14.

Marvin, Dean. 2001. Personal communication from Dean Marvin, The Aerospace Corporation, to Kitt Reinhardt, Air Force Research Laboratory, March 27.

Michaelson, Sol M., and James C. Lin. 1987. Biological Effects and Health Implications of Radiofrequency Radiation. New York: Plenum Press.

Moore, Taylor. 2000. "Renewed Interest in Space Solar Power." *EPRI Journal* 25(1):6-17.

Mullins, Carie. 2000. "Integrated Architecture Assessment Model and Risk Assessment Methodology." Briefing by Carie Mullins, Futron Corporation, to the Committee for the Assessment of NASA's Space Solar Power Investment Strategy, National Academy of Sciences, Washington, D.C., October 23.

Murphy, D., and Allen, D. 1997. "Scarlet Development, Fabrication, and Testing for Deep Space 1 Spacecraft." Pp. 2237-2245 in *Proceedings of the 32st Intersociety Energy Conversion Engineering Conference*, Honolulu, Hawaii.

NASA (National Aeronautics and Space Administration). 1999. *Advanced Space Transportation Program R&T Base Program Plan*. Office of Aero-Space Technology, National Aeronautics and Space Administration, July 11. Available online at <http://astp.msfc.nasa.gov/>. Accessed on August 16, 2001.

NASA. 2000. Information presented at the Workshop on Orbital Debris, Large Scale Space Structures, and Tethers. Marshall Space Flight Center, Huntsville, Alabama, March 14-16.

NIEHS (National Institute of Environmental and Health Sciences). 1999. Health Effects from Exposure to Power-Line Frequency Electric and Magnetic Fields. NIH Publication No. 99-4493. National Institute of Environmental Health Sciences, Research Triangle Park, N.C.: May 4. Available online at <http://www.niehs.nih.gov/emfrapid>. Accessed on August 16, 2001.

NRC (National Research Council), Aeronautics and Space Engineering Board. 1995. *Orbital Debris: A Technical Assessment*. Washington, D.C.: National Academy Press. Available online at <http://books.nap.edu/catalog/4765.html>. Accessed on August 16, 2001.

NSTC (National Science and Technology Council), Committee on Transportation Research and Development. 1995. *Interagency Report on Orbital Debris*. Washington, D.C.: National Science and Technology Council. November.

Olds, John. 2000. "Space Solar Power: Space Transportation and SSP Deployment Support Studies." Briefing by John Olds, Georgia Institute of Technology, to the Committee for the Assessment of NASA's Space Solar Power Investment Strategy, National Academy of Sciences, Washington, D.C., September 14.

Oman, Henry. 2001. "Solar Power from Space." *IEEE AES Systems*, January, pp. 17-26.

O'Neill, Mark, and Mike Piszczor. 2001. *Stretched Lens Array (SLA), Ultralight Concentrator for Space Power.* IECEC Paper No. 2001-AT-39 presented at the 36th International Energy Conversion Engineering Conference, Savannah, Ga., July 31.

Picazo, M.L., E. Martinez, M.V. Carbonell, A. Raya, J.M. Amaya, and J.L. Bardasano. 1999. "Inhibition in the Growth of Thistles (Cynara cardunculus L.) and Lentils (Lens culinaris L.) Due to Chronic Exposure to 50-Hz, 15 µT Electromagnetic Fields." *Electro- and Magnetobiology* 18(2):147-156.

Pignolet, Guy. 1999. "The SPS-2000 'Attaché Case' Demonstrator." *Space Energy and Transpor*tation 4(3,4):125-126.

Pignolet, Guy. 2001. "Recent SSP Activities and Plans in France." Briefing by G. Pignolet, French Space Agency CNES, to the NASA International Forum on Space Solar Power, Washington, D.C., January 12.

Reinhardt, Kitt. 2001a. Air Force Research Laboratory (AFRL) in-house space solar cell modeling studies.

Reinhardt, Kitt. 2001b. Personal communication of Boeing presentation to Air Force in CY 2000 to Committee for the Assessment of NASA's Space Solar Power Investment Strategy, from Kitt Reinhardt, Air Force Research Laboratory, January 30 and April 9.

Skiles, Jay. 2001. Personal communication from Jay W. Skiles, NASA Ames Research Center, to Karen Harwell, National Research Council, August 14.

Sovie, Ronald. 2001. Personal communication from Ronald J. Sovie, Space Directorate, NASA Glenn Research Center to Kitt Reinhardt, Air Force Research Laboratory, January.

Spofford, J.R., and D.L. Akin. 1984. Results of the MIT Beam Assembly Teleoperator and Integrated Control Station, AIAA Paper 84-1890, *Proceedings of the AIAA Guidance and Control Conference,* Seattle, Washington, August.

Urech, Martin, Bernhard Eicher, and Juerg Siegenthaler. 1996. "Effects of Microwave and Radio Frequency Electromagnetic Fields on Lichens." *Bioelectromagnetics* 17(4):327-334.

Whitaker, Chuck. 2001. Personal communication from Chuck Whitaker, Endicon Engineering, to Rhoads Stephenson, retired, Jet Propulsion Laboratory, February 12.

Appendixes

A

Statement of Task

The Aeronautics and Space Engineering Board (ASEB) will assess the technology investment strategy of the "Solar Power from Space" Program to determine its technical soundness and contribution to the roadmap that NASA has developed for this program. The ASEB will assemble a committee with expert knowledge in solar power and associated technologies to conduct an independent technical assessment that will embody the following:

1. Critique the overall technology investment strategy for the Solar Power from Space Program in terms of the plan's likely effectiveness in meeting the program's technical and economic objectives.

2. Identify areas of highest technology investment necessary to create a competitive space-based electric power system.

3. Identify, where possible, opportunities for increased synergy with other research and technology efforts, including the application of these technologies to commercial programs or programs associated with NASA's science and exploration enterprises.

4. Provide, where possible, an independent assessment of the adequacy of available resources for achieving the plan's technology milestones.

5. Recommend changes in the technology investment strategy, as appropriate. In particular, identify gaps or omissions in the program's technology investment strategy that must be filled, if NASA is to field a full-scale system.

The ASEB will draw upon other elements of the NRC, as appropriate, in conducting this study. A final report will be issued at the end of the study.

B

Biographical Sketches of Committee Members

Richard J. Schwartz *(Chair)* has been dean of the Schools of Engineering at Purdue University since July 1995. He has been on the faculty at Purdue since 1964 and has served as a consultant to a number of corporations, both large and small. Dr. Schwartz served as the chairman of the Science and Technology Advisory Committee for the Department of Energy's National Renewable Energy Laboratory. He serves on the Advisory Committee for the National Center for Photovoltaics and has served on the Board of Directors of the National Electrical Engineering Department Heads Association and on the International Committee for the European Union's Photovoltaic Solar Energy Conference. He has served as general chairman of the 23rd Institute of Electrical and Electronic Engineers (IEEE) Photovoltaic Specialists Conference and as a member of the International Committee for the World Conference on Photovoltaic Energy Conversion. In 1987, Dr. Schwartz was named a fellow of the IEEE for his research work on the analysis, design, and development of high-intensity silicon solar cells. In 1998, he received the IEEE William Cherry Award for his contributions to the field of photovoltaics. He received the B.S.E.E. from the University of Wisconsin-Madison in 1957 and the S.M.E.E. and Sc.D. degrees from the Massachusetts Institute of Technology in 1959 and 1962, respectively. While a graduate student, he was one of eight founders of Energy Conversion, Inc., a manufacturer of thermoelectric materials, devices, and systems. He served as vice president of engineering at Energy Conversion, Inc., where he developed new techniques for the growth of single-crystal quaternary thermoelectric materials and high-performance thermoelectric heat pump modules.

Mary L. Bowden is currently visiting professor in the department of Aerospace Engineering at the University of Maryland, affiliated with the Space Systems Laboratory. Her research interests include assembly of structures in extravehicular assembly, large space structures, and the dynamics of space structures. She has been employed in the area of solar array design and material selection by the Able Engineering Company (AEC). While employed by AEC, Dr. Bowden worked in design and test support analysis for deployable structures and other space mechanisms. She has also worked for the American Rocket Company and American Composite Technology in the areas of dynamic structural model development and smart structures. Dr. Bowden graduated with Sc.D. and M.S. degrees from the Massachusetts Institute of Technology (MIT) and a B.A. from Cornell University. She was named Space Educator of the Year in 1995 by the Western Spaceport Technological and Educational Council and awarded a National Aeronautics and Space Administration (NASA) Group Achievement Award for the Experimental Assembly of Structures in EVA (EASE) Flight Experiment. Dr. Bowden was awarded a Zonta Amelia

Earhart Fellowship and a DuPont fellowship during her tenure at MIT.

Hubert P. Davis has been an independent consultant since 1985 performing systems engineering and integration studies. His clients have included the NASA Johnson Space Center, the Large Scale Programs Institute, the University of Texas, United Technologies, the Boeing Company, Rocketdyne, NASA Langley Research Center, and the Lawrence Livermore National Laboratories. In 1980, Mr. Davis founded Eagle Engineering, Inc., in Houston, Texas, a consulting company coupling the experience of Apollo Program leaders with outstanding recent graduates. Throughout the 1970s, Mr. Davis managed Future Programs for the NASA Johnson Space Center, where he developed the Inertial Upper Stage and solid rocket concepts and established the early NASA studies of the space solar power satellite concept. Throughout the 1960s, Mr. Davis had a lead engineering role in the design and development of power and propulsion systems for the Apollo Lunar Landing program. Mr. Davis currently maintains a leadership role in the development of space solar power system concepts.

Richard L. Kline is president of Klintech, a technical consulting company. He is also president and chief executive officer of United Satellite Launch Services, a project to convert Russian missiles to provide scientific research and commercial satellite launch services. Mr. Kline was employed by the Grumman Corporation from 1956 until retiring in 1991. He served as vice president and deputy director, Grumman Space Station Program Support Division. Previously he served as program vice president for civil systems and led Grumman's work in space solar power station concept design. He also initiated and led Grumman's participation in space commercialization. Mr. Kline was employed at NASA, Washington, D.C., from 1992 until 1997 in a number of positions, including directing the Interagency National Facilities Study. He was commended by the Vice President for his contributions to reinventing government and received the NASA Exceptional Achievement Medal for his leadership. In 1997, he joined ANSER as vice president, international activities, and led ANSER's work to promote mutually beneficial scientific and commercial international partnerships in space, primarily with Russia. Mr. Kline has been elected fellow of the American Institute of Aeronautics and Astronautics (AIAA), the American Astronomical Society, the British Royal Aeronautical Society, the British Interplanetary Society, the American Society of Mechanical Engineers, and the Society of Automotive Engineers. He is a licensed professional engineer in New York and Virginia. Mr. Kline is an affiliate professor at George Mason University and is a member of its School for Computational Science Advisory Board. He is a co-chair of the International Astronautical Federation's World Space Congress 2002 Technical Program Committee. Mr. Kline received AIAA's von Braun Space Management Medal and was elected to the International Academy of Astronautics.

Molly K. Macauley is a senior fellow with Resources for the Future (RFF), Washington, D.C. She has been Director of Academic Programs at RFF since 1996. Since 1983, Dr. Macauley's research at RFF has included the areas of public finance, energy economics, regulation of toxic substances, environmental economics, advanced materials economics, the value of information, and economics and policy issues of outer space. Dr. Macauley's space research includes the valuation of nonpriced space resources, the design of incentive arrangements to improve space resource use, and the appropriate relationship between public and private endeavors in space research, development, and commercial enterprise. Dr. Macauley has been a visiting professor at Johns Hopkins University, Department of Economics, and at Princeton University, Woodrow Wilson School of Public Affairs. Dr. Macauley testified before Congress on the Commercial Space Act of 1997, the Omnibus Space Commercialization Act of 1996, the Space Business Incentives Act of 1996, and space commercialization. Dr. Macauley has served on many national-level committees and panels, including the congressionally mandated Economic Study of Space Solar Power (chair); the National Research Council's (NRC's) Board on Physics and Astronomy, Helium Reserve Committee; the NRC Space Studies Board Steering Group on Space Applications and Commercialization; and the NRC Space Studies Board Task Force on Priorities in Space Research. Dr. Macauley has published extensively over the past 16 years, with more than 70 journal articles, books, and chapters of books. Dr. Macauley serves on the Board of Directors of Women in Aerospace and has served as president of the Thomas Jefferson Public Policy Program, College of William and Mary.

Lee D. Peterson is an associate professor of aerospace engineering sciences at the University of Colorado, Boulder. He has been an associate professor or assistant professor at the University of Colorado since 1991. Dr. Peterson is also director of the McDonnell-Douglas Aerospace Structural Dynamics and Control Laboratory and is a member of the multidisciplinary Center for Aerospace Structures. His principal area of research is in high-precision deployable spacecraft structures for use in optical telescopes and interferometers. His research group has experimentally characterized and modeled a new class of nonlinear mechanics that limits the stability of such space structures at nanometer levels of motion. He has also made research contributions in experimental structural dynamics, system identification, parameter identification joint modeling, and active structural control. Dr. Peterson is also actively involved in the University of Colorado's new undergraduate aerospace curriculum and served as the technical director of the Integrated Teaching and Learning Laboratory from 1995 to 1997. From 1989 to 1991, Dr. Peterson was assistant professor of Aeronautics and Astronautics at Purdue University. From 1987 to 1989, he was a member of the technical staff at Sandia National Laboratories, Albuquerque, New Mexico.

Kitt C. Reinhardt is an electrical engineer conducting photovoltaic device research and development in the Space Vehicles Directorate of the U.S. Air Force Research Laboratory, Albuquerque, New Mexico. Dr. Reinhardt was the Air Force nominee and the year 2000 winner of the Rotary National Award for Space Achievement, an early career award based on Dr. Reinhardt's pioneering work in the development of high-efficiency, multijunction solar cells as well as ultralightweight flexible thin-film photovoltaics for next-generation space systems. Dr. Reinhardt led the successful development and commercialization of the first 25 percent-efficient space solar cell, as well as the invention and current development of the first 30-35 percent efficient space solar cell. In addition, he has been instrumental in several revolutionary areas, such as thin-film photovoltaics and advanced thermal-to-electric conversion. Most recently, Dr. Reinhardt, together with Hong Hou from Sandia National Laboratories, invented an entirely new approach capable of achieving 35-40 percent solar-to-electric conversion with a four-junction solar cell design. A patent for the device was granted in August 1999.

R. Rhoads (Rody) Stephenson retired from the Jet Propulsion Laboratory (JPL) in 1998, where he had been deputy director of the JPL technology program since 1991 and acting director since 1995. The technology program included all of JPL's technology development efforts, including robotics and its space power work. In this capacity, Dr. Stephenson was involved in many studies of space power beaming to Earth. He also worked, in conjunction with Langley Research Center, on large space structures and the JPL program on control-structures interaction, providing a technology base for the space interferometer project. Between 1981 and 1991, Dr. Stephenson was manager of the Electronics and Control Division at JPL. The division included the power section, which had responsibility for all forms of space power, including solar power, and it participated in solar cell development and testing and in the solar power beam transmission studies of that period. Most recently in his 36-year career at JPL, the laboratory turned to Dr. Stephenson to serve as a member of the Galileo and Cassini Review Boards, to chair the Mars Pathfinder Board, and to lead the internal failure Review Board for the Mars Observer mission.

Dava Newman, Aeronautics and Space Engineering Board liaison to the Committee for the Assessment of NASA's Space Solar Power Technical Investment Strategy, is an associate professor of aeronautics and astronautics at the Massachusetts Institute of Technology and a MacVicar faculty fellow. She conducts multidisciplinary efforts combining aerospace bioengineering, human-in-the-loop dynamics and control modeling, biomechanics, human interface technology, life sciences, and systems analysis and design. Dr. Newman served as a member of the NRC Committee on Advanced Technology for Human Support in Space and the Committee on Engineering Challenges to the Long-Term Operation of the International Space Station.

C

Example of NASA's SERT Program Technology Roadmaps

Figures C-1, C-2, and C-3 are examples of the roadmaps and programmatic charts for each individual technology area in NASA's SERT program. The figures are presented in original, unedited form.

REFERENCE

Mankins, John and Joe Howell. 2000. "Strategic Research and Technology Road Map." Briefing by John Mankins and Joe Howell, National Aeronautics and Space Administration, to the Committee for the Assessment of NASA's Space Solar Power Investment Strategy, National Research Council, Washington, D.C., December 14.

Solar Power Generation
Research & Technology Executive Summary

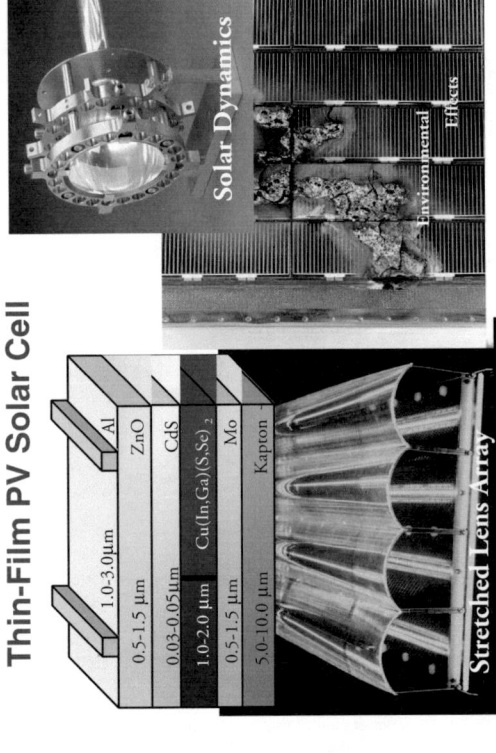

Thin-Film PV Solar Cell

Stretched Lens Array

DESCRIPTION
- Dramatic advances in solar power generation (SPG) are enabling for large-scale SSP and electric propulsion.
- SPG R&D is needed to achieve large (0.1 - 1.0 MW Class and higher) collection of solar energy and conversion to electrical power in systems with very low cost and mass

APPROACH
- A portfolio of investments, guided by studies, including long-term, high-risk technology research and demonstrations focused on enabling interim applications of SPG technology.
- Harvesting existing investments in near term demonstrations, while pursuing multiple in-house, contracted and leverage R&T paths for large SPG systems

PARTICIPANTS
- NASA Field Centers (e.g., GRC, JPL, MSFC); Other Agency Labs (e.g.,NREL, AFRL, NRL); Industry (e.g., UAT Boeing, Entech, ER, SRS); Universities (e.g., Fisk Univ., Auburn)

TECHNOLOGY ELEMENTS/CHALLENGES
- Multi-Band Gap (MGB) Cells
  - Stacked and Rainbow
- Thin Film PV
  - Inorganic-Organic low cost PV cell; Deployable low mass structures
- Quantum Well / Quantum Dot
- Large array development / deployment techniques
- Concentrators
  - PV and solar thermal dynamic
- Solar Dynamics
  - Heat Engines, radiators
- Environmental Interactions
  - Arc mitigation, high voltage operations, etc.

OBJECTIVES
- Stretched Lens Array: 30% conc. cell/27.4% panel @378 W/kg
- 15% thin film cell on flexible substrate
- SLA-Rainbow 45% concentrator array demo
- Advanced Concepts (e.g., Quantum Dot Cell, 20m Inflatable solar electric sail
- 50 kW Electric Propulsion/Generation Flight Test

BENEFITS
- SEPS for science missions
- High power for HEDS missions (transport and surface)
- Affordable high power for ïnew space industriesî

FIGURE C-1 Sample SERT program executive summary chart on solar power generation activity. SOURCE: Mankins and Howell, 2000.

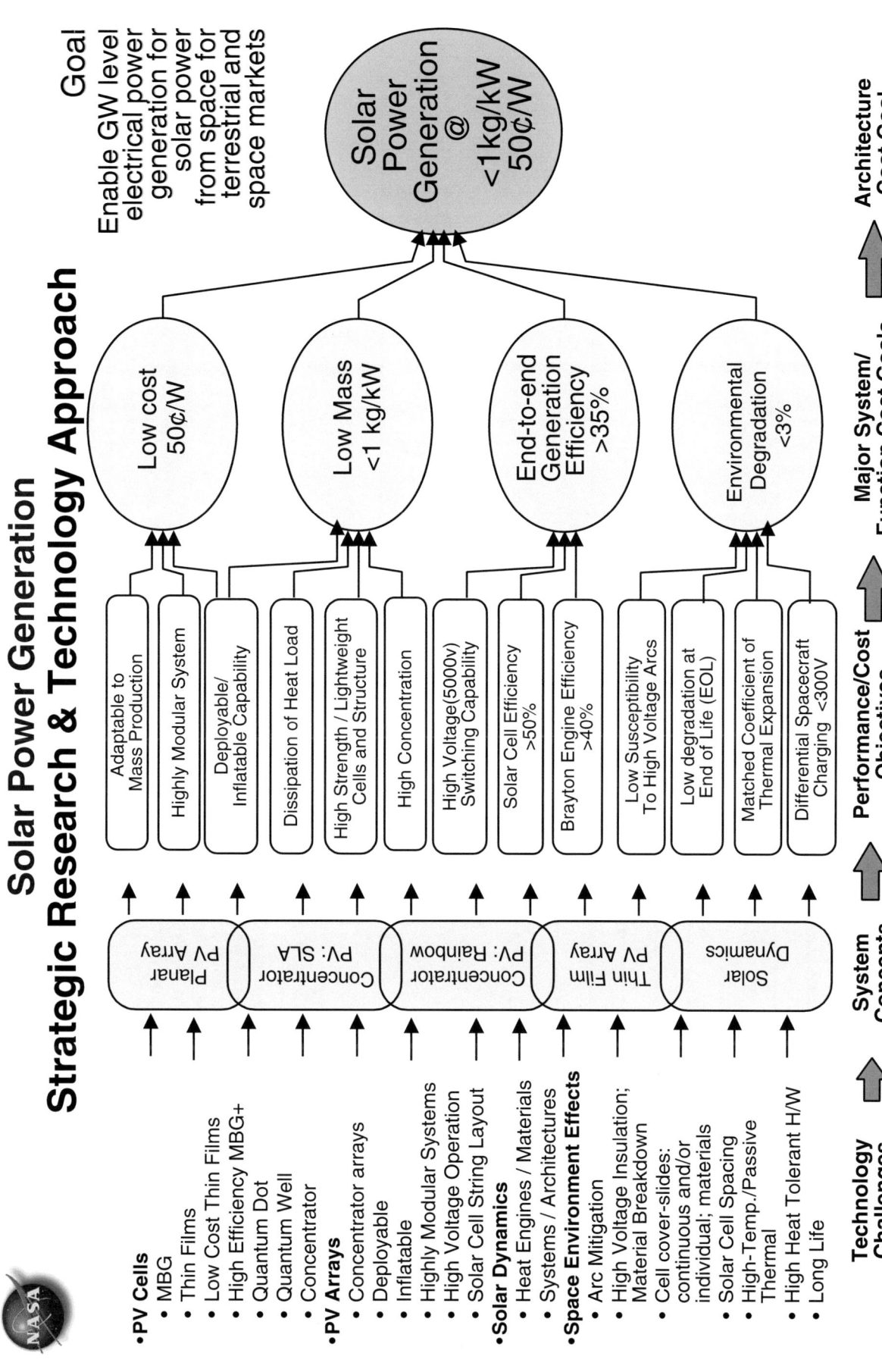

FIGURE C-2 Sample SERT program goal chart on solar power generation activity. SOURCE: Mankins and Howell, 2000.

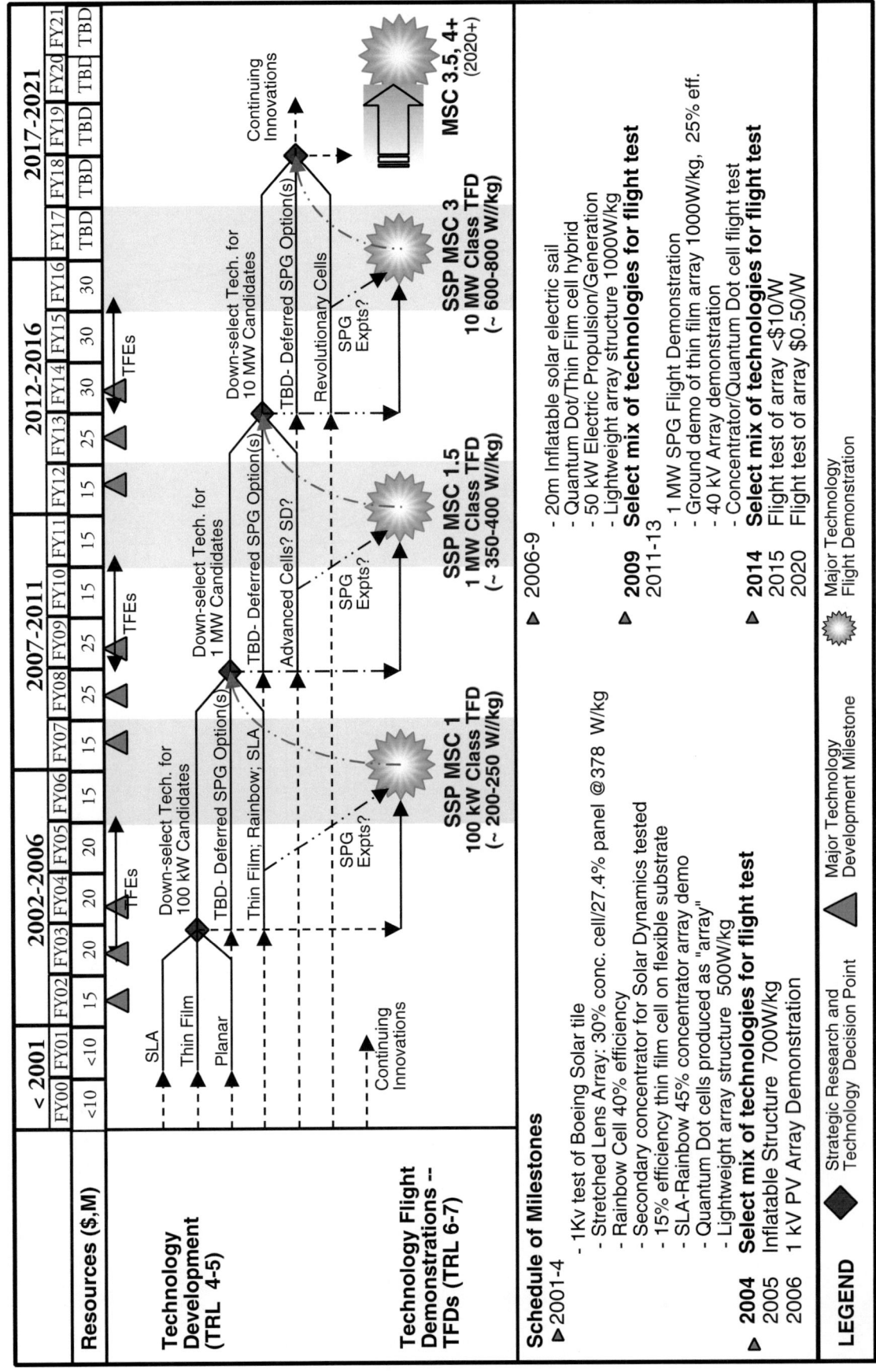

FIGURE C-3 Sample SERT program milestones chart on solar power generation activity. SOURCE: Mankins and Howell, 2000.

D

Brief Overview of NASA's Space Solar Power Program

BASELINE SSP SYSTEMS

A space solar power system requires integration of many technologies in order to generate electricity from the Sun. Figure D-1 depicts a generic solar power system, from collection of the solar power to receipt of the solar power on Earth and delivery to the grid. This concept is based on the use of microwaves. Options using lasers would involve constellations of small individual satellites, each with its own transmitter, as depicted in Figure D-2. The National Aeronautics and Space Administration's (NASA's) Space Solar Power (SSP) Exploratory Research and Technology (SERT) program, as of the date of this report, has not yet chosen a baseline system. Several possible variations of flight demonstrations and systems have been presented to the committee, each classified according to four model system categories (MSCs). Refer to Section 2-1 for a more detailed description of these demonstrations and program milestones.

Despite the differences in these concepts, all space solar power systems have a set of common technology areas and work in the same general manner. Solar energy is collected in geosynchronous Earth orbit (GEO) by a solar power generation technology, probably consisting of photovoltaic (PV) arrays that capture radiation from the Sun and convert it (using the photovoltaic process) into direct electric current. These PV arrays blanket a surface that faces the Sun at all times. The electric current is collected and transformed through the power management and distribution system. Transmitters then beam the power via wireless power transmission to a specific collector (either on Earth's surface or in space). Receivers (on Earth's surface or in space) collect the incoming microwave or laser transmission energy and convert it into electricity. For microwave systems, this collector is referred to as a rectenna. For laser-based transmission, the collector is constructed from solar arrays. For space-to-space systems, the collector is application specific. The construction of such SSP systems, each on the order of several square kilometers in size, is handled almost entirely through autonomous robotic assembly, inspection, and maintenance in GEO and requires numerous launches of heavy payloads into space. In-space transportation of SSP components is also required to move payloads from low Earth orbit to GEO. Various risk management and systems design tools also need to be developed during the design stages of any SSP system.

OVERVIEW OF NASA'S SPACE SOLAR POWER (SSP) EXPLORATORY RESEARCH AND TECHNOLOGY (SERT) PROGRAM

NASA's SERT program mainly involves research on technologies and design methods that is necessary for such a huge undertaking. The program has identi-

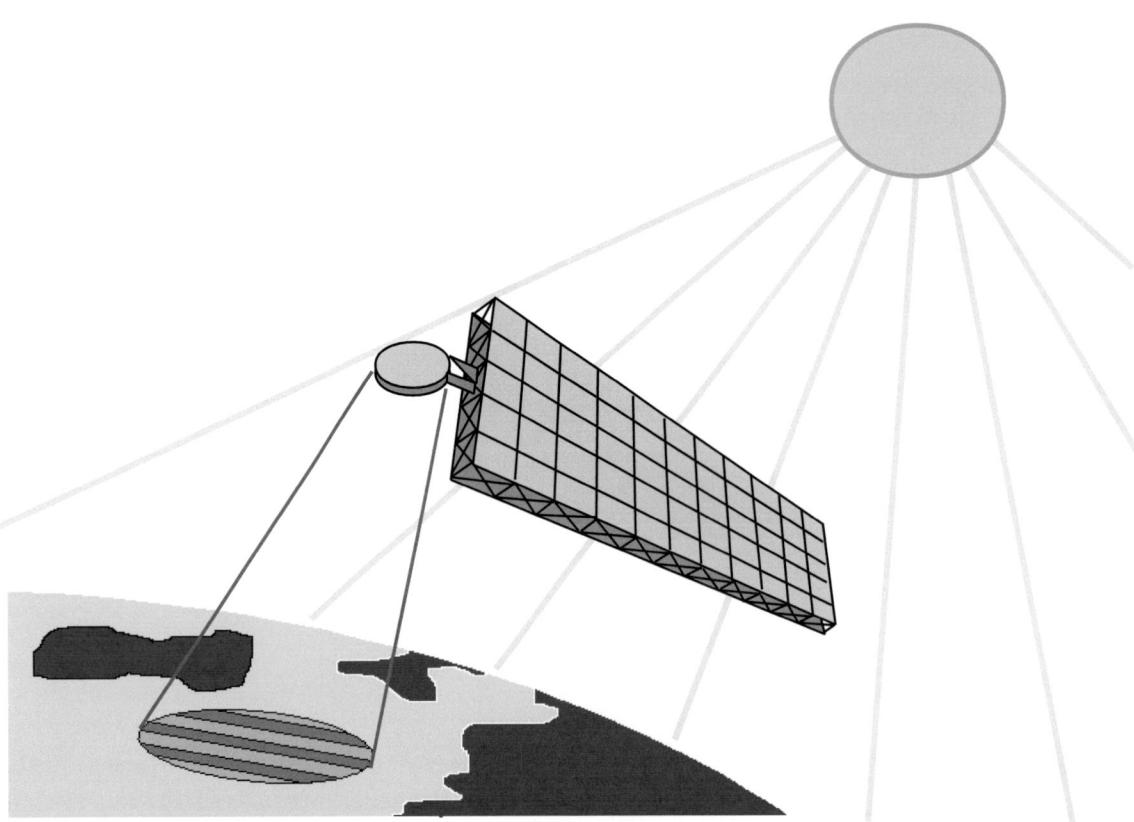

FIGURE D-1 Generic space solar power system. SOURCE: Adapted in part from Nansen, 2000.

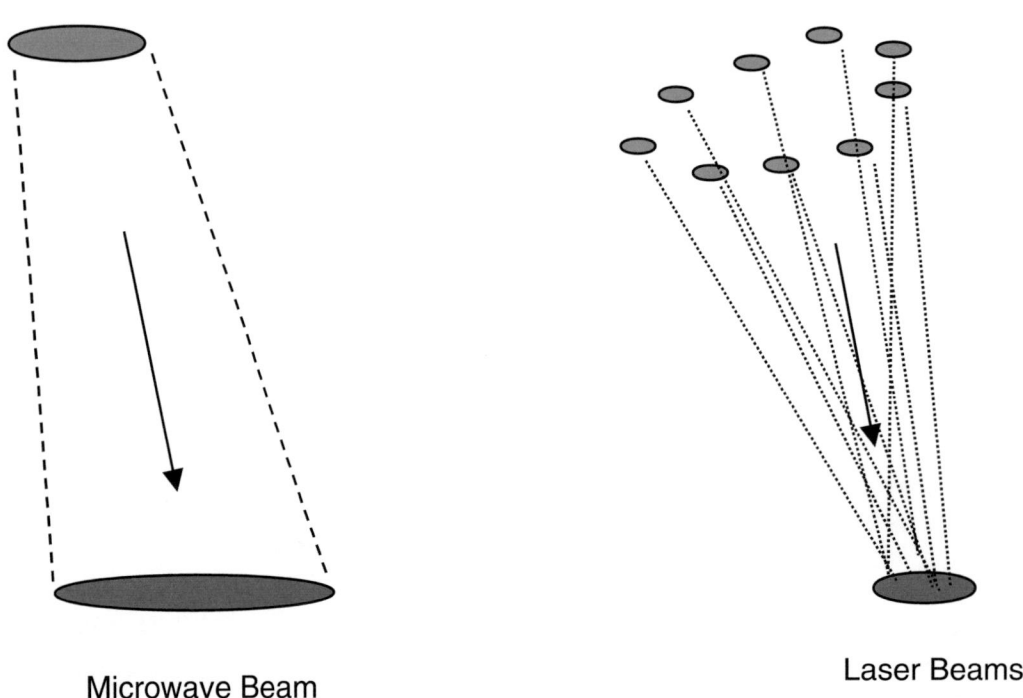

FIGURE D-2 Generic microwave and laser SSP systems. SOURCE: Adapted in part from Dickinson, 2000.

fied several flight demonstration milestones in order to test technologies and concepts in the near-term and mid-term in preparation for transferring the technologies to industry for final full-scale development and implementation. A more specific treatment of these flight demonstrations and key program milestones can be found in Section 2-1.

NASA has chosen to break its research into 12 areas for funding:

1. Systems integration, analysis, and management
2. Solar power generation
3. Wireless power transmission
4. Space power management and distribution
5. Structural concepts, materials, and controls
6. Thermal management and materials
7. Space assembly, inspection, and maintenance
8. Platform systems
9. Ground power systems (GPS)
10. Space transportation (Earth-to-orbit and in-space)
11. Environmental, health, and safety
12. Economic analysis

Each area (with the exception of economic analysis) has been allocated a portion of the earmarked government funding provided to the SERT program for technology roadmap development and prioritization and was charged with (1) developing a set of cost and technology goals, (2) compiling a list of important technology challenges, (3) developing potential applications of technology advancements, (4) developing a breakdown of the specific work necessary for advancement, and (5) developing a schedule of technology milestones that parallel the milestones of the total program. An example of these roadmaps and goals for the solar power generation portion of the program can be found in Appendix C. The program has identified an investment portfolio for a future SSP program with planned resource allocation through 2016 (see Table D-1). This allocation will be affected by choices made by NASA and the President's Office of Management and Budget in space solar power. Technology flight demonstrations (referred to by NASA as MSCs) are scheduled in FY 2006-2007, FY 2011-2012, and FY 2016.

The SERT program has several levels of organization stemming from management at the NASA Office of Space Flight. A schematic of this organizational structure, which incorporates many NASA field centers as well as industry and academia, is shown in Chapter 3, Figure 3-1. The program has created several levels of oversight through its Senior Management Oversight Committee and various technical and systems working groups. The program has also obtained various external evaluations from groups such as the National Research Council; Resources for the Future, an economic research group; and professional technical societies such as the American Institute of Aeronautics and Astronautics. External comment has also been provided through involvement in various international organizations and symposiums such as the International Forum on Space Solar Power.

REFERENCES

Dickinson, Richard. 2000. "Wireless Power Transmission." Briefing by Richard Dickinson, Jet Propulsion Laboratory, to the Committee for the Assessment of NASA's Space Solar Power Investment Strategy, National Academy of Sciences, Washington, D.C., September 13.

Mankins, John and Joe Howell. 2000. "Strategic Research and Technology Roadmap." Briefing by John Mankins and Joe Howell, National Aeronautics and Space Administration, to the Committee for the Assessment of NASA's Space Solar Power Investment Strategy, National Academy of Sciences, Washington, D.C., December 14.

Nansen, Ralph. 2000. "The Space Solar Power Solution: An Industry/Government Partnership." Briefing by Ralph Nansen, Solar Space Industries, to the Committee for the Assessment of NASA's Space Solar Power Investment Strategy, National Academy of Sciences, Washington, D.C., October 23.

TABLE D-1 Proposed Space Solar Power Program Resources Allocation, FY 2000 to FY 2016 (millions of dollars)

Funding Area	FY00	FY01	FY02	FY03	FY04	FY05	MSC 1 FY06	FY07	FY08	FY09	FY10	MSC 1.5 FY11	FY12	FY13	FY14	FY15	MSC 3 FY16
Systems integration and management	4	4	5	7	8	8	8	10	10	10	10	10	10	10	10	10	10
Solar power generation	10	10	15	20	20	20	15	15	25	25	15	15	15	25	30	30	30
Wireless power transmission	5	5	8	10	15	15	25	30	40	40	45	60	35	30	35	40	40
Power management and distribution	5	5	7	10	15	15	10	10	15	15	15	10	10	20	25	30	25
Structural concepts, materials, and controls	10	10	10	10	15	20	20	30	50	50	50	45	35	30	30	35	35
Thermal materials and management	1	1	5	7	15	20	20	20	20	25	30	30	30	30	30	30	30
Space assembly, inspection, and maintenance	0.01	0.01	10	15	20	25	30	30	30	30	30	30	30	35	40	40	40
Platform systems	1	1	2	3	4	4	4	4	4	4	4	4	4	4	4	4	4
Ground power systems	1	1	2	2	3	4	4	5	5	5	10	10	10	10	10	10	10
Earth-to-orbit transportation and infrastructure	0.1	0.1	1	1	1	1	1	1	1	1	1	1	1	1	1	1	1
In-space transportation and infrastructure	5	5	10	10	15	20	20	20	15	15	20	20	20	15	15	20	20
Environmental, health, and safety factors	1	1	3	4	5	5	5	5	5	5	5	5	5	5	5	5	5
Technology flight demonstrations	1	1	10	25	75	125	150	200	250	350	500	500	500	650	750	750	750
Total	44.11	44.11	88	124	211	282	312	380	470	575	735	740	705	865	985	1,005	1,000

SOURCE: Adapted in part from Mankins and Howell, 2000.

E

Participants in Committee Meetings

The full committee met four times between September 2000 and March 2001. Outside participants are listed below, grouped by organization:

Auburn University
Henry Brandhorst

Canadian Space Agency
Bryan Erb

Carnegie Mellon University
Sarjoun Skaff
Peter Steritz

Dow Jones Newswires
Bryan Lee

Futron Corporation
Carie Mullins

Georgia Institute of Technology
John Olds

NASA Ames Research Center
Charles Neveu
Hans Thomas

NASA Glenn Research Center
Sheila Bailey
James Dolce
James Dudenhoefer
Lee Mason
Barbara McKissock
Richard Shaltens

NASA Headquarters
John Mankins

NASA Jet Propulsion Laboratory
Richard Dickinson
Brad Kennedy
Neville Marzwell
David Maynard

NASA Johnson Space Center
Christopher Culbert

NASA Langley Research Center
Chris Moore

NASA Marshall Space Flight Center
Jeffrey Anderson
Connie Carrington
Daniel Davis
Joe Howell

President's Office of Management and Budget
Brant Sponberg

Pacific Northwest National Laboratory
James Dooley

Science Applications International Corporation
Harvey Feingold

Solar Space Industries
Ralph Nansen

Strategic Insight, Ltd.
John Fini

Sunsat Energy Council
Frederick Koomanoff

F

Acronyms and Abbreviations

AC	alternating current		g	gravity level at Earth's surface
AFRL	Air Force Research Laboratory		GEO	geosynchronous Earth orbit
AIAA	American Institute of Aeronautics and Astronautics		GHz	gigahertz
			GPS	ground power system
a-Si	amorphous silicon		GW	gigawatt
AST	Advanced Space Transportation (program)		HEDS	Human Exploration and Development of Space (NASA)
B	billion		in.	inch
CIGS	copper indium gallium diselenide		IS	intelligent system
cm	centimeter		ISS	International Space Station
CNES	Centre National d'Études Spatiales (French Space Agency)		ISTI	in-space transportation and infrastructure
			ITAM	integrated technology analysis methodology
DC	direct current			
DOD	Department of Defense		kg	kilogram
DOE	Department of Energy		km	kilometer
			kW	kilowatt
EH&S	environmental health and safety		kW-hr	kilowatt-hour
EOL	end of life			
EPRI	Electric Power Research Institute		LEO	low Earth orbit
ESA	European Space Agency			
ETI	Earth-to-orbit transportation and infrastructure		M	million
			m^2	square meter
ETO	Earth to orbit		m^3	cubic meter
			MBG	muti-band gap
ft	foot		MHD	magnetohydrodynamics

μm	micrometer	SCMC	structural concepts, materials, and controls
MSC	model system category	SERT	SSP Exploratory Research and Technology (program)
mW	milli-watt		
MW	megawatt		
		SIM	systems integration and management
NASA	National Aeronautics and Space Administration	SLA	stretched lens array
		SLI	Space Launch Initiative
NRC	National Research Council	SMOC	Senior Management Oversight Committee
NREL	National Renewable Energy Laboratory	SPG	solar power generation
NRO	National Reconnaissance Office	SPMAD	space power management and distribution
		SPS	solar power satellite
OTA	Office of Technology Assessment	SSP	space solar power
PMAD	power management and distribution	TFD	technology flight demonstration
PS	platform systems	TMM	thermal materials and management
PV	photovoltaic	TRL	technology readiness level
PVMat	PV Manufacturing Technology	TSX	Tele-robotic Shuttle Experiment
R&D	research and development	USAF	U.S. Air Force
R&T	research and technology		
RMS	Remote Manipulator System	W	watt
		WPT	wireless power transmission
SAIM	space assembly, inspection, and maintenance	YAG	yttrium aluminum garnet
		yr	year